Pocket BIRDS of CANADA

10700533

CONSULTANT EDITOR
DAVID M. BIRD, PH.D.
Emeritus Professor of Wildlife Biology
McGill University

Illustrator	**Editor**
John Cox,	Jamie Ambrose, Lori Baird,
Andrew Mackay	Tamlyn Calitz, Marcus Hardy,
Jacket Design	Patrick Newman, Siobhan O'Connor,
Development Manager	David Summers, Miezan van Zyl,
Sophia MTT	Rebecca Warren
Picture Editor	**Jacket Editor**
Neil Fletcher	Claire Gell
Picture Researchers	**Map Editor**
Laura Barwick, Will Jones	Paul Lehman
Producer	**Managing Editor**
Mary Slater	Barbara Campbell
Publisher	**Publishing Director**
Liz Wheeler	Jonathan Metcalf

DK DELHI

Art Editor	**Editor**
Divya P R	Priyanka Kharbanda
Assistant Art Editors	**Managing Editor**
Anusree Saha, Sonakshi Singh	Rohan Sinha
Jacket Designer	**Managing Jackets Editor**
Surabhi Wadhwa	Saloni Singh
Managing Art Editor	**Pre-Production Manager**
Sudakshina Basu	Balwant Singh
Senior DTP Designer	**Production Manager**
Harish Aggarwal	Pankaj Sharma
DTP Designers	
Dheeraj Singh,	
Syed Mohammad Farhan	

First Edition, 2016
DK Canada
320 Front Street West, Suite 1400
Toronto, Ontario M5V 3B6

Library and Archives Canada Cataloguing in Publication
Pocket birds of Canada / consultant editor,
David M. Bird, PhD.
Includes index.
ISBN 978-1-55363-267-2 (paperback)
1. Birds--Canada--Identification.
2. Bird watching--Canada--Guidebooks.
I. Bird, David M. (David Michael), 1949-, editor
QL685.P63 2016   598.0971   C2015-906128-8

DK books are available at special discounts when purchased in bulk for corporate
sales, sales promotions, premiums, fund-raising, or educational use.
For details, please contact specialmarkets@dk.com.

Printed and bound in China

A WORLD OF IDEAS:
SEE ALL THERE IS TO KNOW

www.dk.com

CONTENTS

THIS PAGE: Peregrine Falcon

TITLE PAGE: Gray Jay

How this book works

This guide covers 434 Canadian bird species, which are organized into chapters of related family groupings. Within each chapter, the birds are arranged broadly by genus, so that similar-looking species appear together for ease of comparison. The main index lists the common and scientific names of each bird, and a quick index on the inside back cover provides a handy reference for general bird names.

▽ **INTRODUCTION**
Each chapter opens with an introductory page, briefly describing each family's shared characteristics.

Birds of Prey

Only one species of vulture occurs in Canada: the Turkey Vulture, whose acute sense of smell enables it to detect carrion hidden from sight beneath the forest canopy. This vulture can stay in the air for hours on end, using lift provided by updrafts.

Falcons range in size from the diminutive American Kestrel to the large, powerful Gyrfalcon. They also include the Merlin, the Prairie Falcon, and the fast-diving Peregrine Falcon. Falcon prey ranges from insects to large hare-sized mammals and birds.

Eagle and hawk species cover a wide range of raptors of varying sizes and hunting methods. Forest-dwelling hawks rely on speed and stealth to pounce on small birds among the trees, while the Osprey hovers over water until it sees a fish below, then dives to pluck its prey out of the water with its talons.

COMMON NAME

SCIENTIFIC NAME

DESCRIPTION
Conveys the main features and essential characteristics of the species; may include interesting facts or notable behaviors.

PHOTOGRAPHS
Illustrate the bird in different views, sexes, or plumage variations. Unless otherwise stated, the bird shown is an adult.

194 WAXWINGS &

Bohemian

Bombycilla garrulus

The Bohemian Waxwing species in North Amer Canada. The species is movement is notorious of wild fruits. They are with distinctive rusty u

wispy crest

gray upperparts

black throa

♀

yellow tail band

chestnut undertail feathers

ornat wing mark

▷ **SINGLE-PAGE ENTRIES**
More commonly seen species are given a full-page entry, often showing more photos with varieties of age, sex, and plumage.

Magnolia Warbler
Dendroica magnolia

The bold, flashy, and common Magnolia Warbler is hard to miss as it flits around at eye level, fanning its uniquely marked tail. This species nests in young forests and woders in almost any habitat, so its numbers have not suffered as recent decades, unlike some of its relatives. Although it rarely has no preference for its particular plant. The 19th-century ornithologist Alexander Wilson discovered one feeding in a magnolia tree during migration, which is how it got its name.

Mountain

Poecile gambeli

The Mountain Chickad other chickadees, it stor defend their winter terr when seeds are scarce. E eyebrow and buff-tinge

EASTE

VOICE Raspy tsick-jee-je descending notes bee-bee-ba
NESTING Natural tree cavity woodpecker hole, lined with eggs; 1–2 broods; May–Jun.
FEEDING Forages high in tr spiders; seeds and berries; in preparation for winter.
HABITAT High elevation con may move to foothills and val
LENGTH 5¼in (13.5cm)
WINGSPAN 8½in (22cm)

OTHER KEY INFORMATION
VOICE: *a description of the bird's calls and songs.*
NESTING: *type of nest and its usual location; number of eggs in a clutch; number of broods in a year; breeding season.*
FEEDING: *how, where, and what the bird feeds upon.*
HABITAT: *a description of the bird's preferred habitats in Canada.*
LENGTH AND WINGSPAN: *length is tip of tail to tip of bill; measurements are averages or ranges.*

▽ SPECIES ENTRIES
The typical page describes two bird species. Each profile follows the same easy-to-access structure and features photographs taken in the bird's natural setting.

STATUS
The conservation status of the species, based loosely on the Canadian Wildlife Species at Risk list. Some species are given two statuses, referring to different populations.

S	Stable	**T**	Threatened
D	Declining	●	Endangered

ICKADEES

Waxwing   ⑤

the wilder and rarer of the two waxwing breeds mainly in Alaska and western atory, but the extent of its wintertime ariable, depending on the availability htly larger than Cedar Waxwings tail feathers.

FLIGHT ILLUSTRATION
Shows the bird in flight, from above or below—differences of season, age, or sex are not always visible.

ariable
rest

yellow edges
to outer flight
feathers

gray-brown
upperparts

♂

gray
underparts

VOICE *Dull trill; flocks vocalize constantly with remarkable effect.*
NESTING *Dishevelled cup of sticks and grasses, placed in tree; 4–6 eggs; number of broods unknown; Jun–Jul.*
FEEDING *Insects on the wing in summer; berries of native and exotic trees and shrubs.*
HABITAT *Coniferous forests near disturbed areas while breeding; forest edges, hedges, and residential areas in winter.*
LENGTH *8½in (21cm)*
WINGSPAN *14½in (37cm)*

MAPS
Each profile includes a map showing the range of the bird, focusing on within Canada, with colors indicating seasonal movements.

■	Resident all year
▨	Summer distribution
▨	Winter distribution
▨	Seen on migration

Chickadee   ⑤

s found at elevations of up to 12,000ft (3,600m). Like pine and spruce seeds for harsh winters. Social groups ries and food resources, migrating to lower elevations ls in the Rocky Mountains have a conspicuous white anks.

white
eyebrow

black
crown

white
cheeks

dull gray
upperparts

black
bib

gray
flanks

short,
black
bill

pale gray
underparts

WESTERN

whistled

old
ss and fur; 7–9

for insects and
es seeds in fall

ous forests;
es in winter.

buff-tinged
flanks

EASTERN

SYMBOLS
Indicate sex, age, or season. If no symbols are present, it means that the species exhibits no significant differences in these.

♀ female		♂ male	
🐦 adult		🐦 immature	🐦 juvenile
🌱 spring		☼ summer	🍂 autumn
❄ winter			

Anatomy

In spite of their external diversity, birds are remarkably similar internally. For birds to be able to fly, they need light and rigid bones, a lightweight skull, and hollow wing and leg bones. In addition, pouch-like air sacs are connected to hollow bones, which reduce a bird's weight. The breast muscles, crucial for flight, attach to the keeled sternum (breastbone).

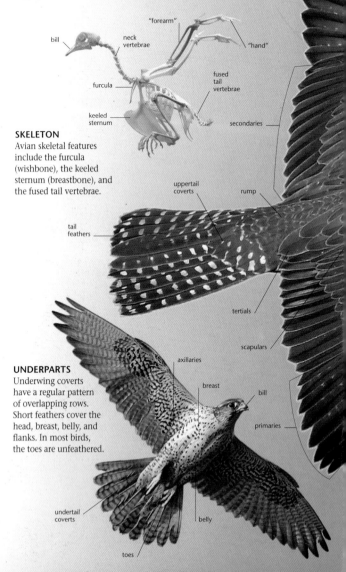

bill

neck vertebrae

"forearm"

"hand"

furcula

fused tail vertebrae

keeled sternum

secondaries

SKELETON
Avian skeletal features include the furcula (wishbone), the keeled sternum (breastbone), and the fused tail vertebrae.

uppertail coverts

rump

tail feathers

tertials

scapulars

axillaries

breast

bill

UNDERPARTS
Underwing coverts have a regular pattern of overlapping rows. Short feathers cover the head, breast, belly, and flanks. In most birds, the toes are unfeathered.

primaries

undertail coverts

belly

toes

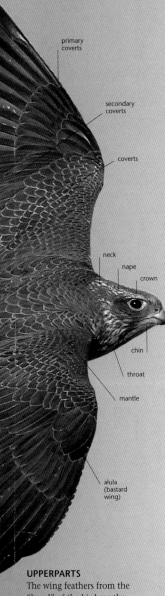

primary
coverts

secondary
coverts

coverts

neck

nape

crown

chin

throat

mantle

alula
(bastard
wing)

Feathers

Feathers serve two main functions: insulation and flight. Small down feathers form an insulating underlayer, and are also the first feathers that nestlings have after hatching. Contour feathers cover the head and body. The rigidity of the flight feathers helps to create a supporting surface that birds use to generate thrust and lift.

FLIGHT FEATHER **CONTOUR FEATHER** **DOWN FEATHER**

Feet and toes

When we talk about a bird's feet we really mean its toes. The structure of the foot can give clues to a bird's characteristics and habits.

WALKING
Ground-foraging birds usually have a long hind claw.

CLIMBING
Most climbers have two toes forward and two backward.

SWIMMING
Water-loving birds have webbing between their toes.

HUNTING
Birds of prey have powerful toes and strong, sharp claws.

UPPERPARTS

The wing feathers from the "hand" of the bird are the primaries and those on the "forearm" are the secondaries. Each set has its accompanying row of coverts. The tertials are adjacent to the secondaries.

Identification

Some species are easy to identify, but in many cases, species
identification is tricky. In Canada, a notoriously difficult group in
terms of identification is the wood-warblers, especially in the fall,
when most species have similar greenish or yellowish plumage.

Geographic range

Each species of bird in Canada lives in a particular area that is called its
geographic range. Some species have a restricted range; others are found
right across Canada and parts south. Species with a broad range usually
breed in a variety of vegetation types, while species with narrow ranges
often have a specialized habitat.

BROAD RANGE
*Red-tailed Hawks range from coast to coast in
North America and south down to Mexico, and
are found in a wide variety of habitats.*

RESTRICTED RANGE
*Whooping Cranes breed only in Wood
Buffalo National Park in Alberta and the
Northwest Territories.*

Size

From hummingbird to Tundra Swan, such is the range of sizes and weights
found among the bird species of Canada. Size can be measured in several
ways, including the length of a bird from bill-tip to tail-tip, or its wingspan.
Comparing the sizes of two birds can also be helpful: for example, the
less familiar Bicknells' Thrush can be compared with the well-known
American Robin.

SIZE MATTERS
*Smaller shorebirds with short legs and
bills forage in shallow water, and larger
ones with longer legs and bills can feed
in deeper water.*

SEMIPALMATED SANDPIPER

LONG-BILLED CURLEW

General shape

Bird body shapes vary widely and can give clues to the habitat in which they live. The American Bittern's long, thin body blends in with the reed beds that it favors. The round-bodied Sedge Wren hops in shrubby vegetation or near the ground where slimness is not an advantage.

SEDGE WREN

AMERICAN BITTERN

Bill shape

In general, bill form, including length or thickness, corresponds to the kinds of food a bird consumes. With its pointed bill, the Mountain Chickadee picks tiny insects from crevices in tree bark. At another extreme, dowitchers probe mud with their long thin bills, feeling for worms.

AMERICAN ROBIN — worms and fruit

CHICKADEE — tiny insects, seeds

HOUSE FINCH — seeds and caterpillars

LONG-BILLED DOWITCHER — worms from deep mud

AMERICAN AVOCET — small shrimps in water

GREAT BLUE HERON — fish

GOLDEN EAGLE — mammals and birds

SURF SCOTER — marine mollusks

Wing shape

Birds' wing shapes are correlated with their flight style. The long, broad wings of the Red-tailed Hawk are perfect for soaring, while the tiny wings of hummingbirds allow them to hover in front of flowers for a meal of nectar.

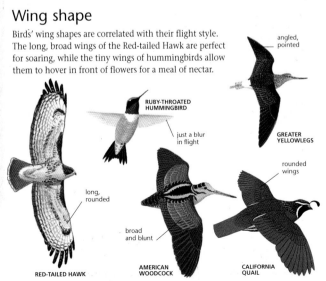

RUBY-THROATED HUMMINGBIRD — just a blur in flight

GREATER YELLOWLEGS — angled, pointed

RED-TAILED HAWK — long, rounded

AMERICAN WOODCOCK — broad and blunt

CALIFORNIA QUAIL — rounded wings

Tail shape

Tail shapes vary as much as wing shapes, but are not so easily linked to a function. Irrespective of shape, tails are needed for balance. In some birds, tail shape, color, and pattern are used in courtship displays or in defensive displays when threatened.

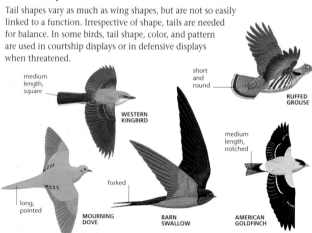

medium length, square

WESTERN KINGBIRD

short and round

RUFFED GROUSE

medium length, notched

long, pointed

MOURNING DOVE

forked

BARN SWALLOW

AMERICAN GOLDFINCH

Colors and markings

Colors and markings, including stripes, bars, patches, and spots, can be instrumental in identifying a species. Look for wing and tail patterns, which may be noticeable only when the bird opens its wings or spreads its tail, or distinctive patterns on the head.

black spots

LAZULI BUNTING

black-and-white streaks

BLACK AND WHITE WARBLER

black-and-white head pattern

WHITE-CROWNED SPARROW

WOOD THRUSH

streaking on belly

BARRED OWL

white wing bars

white eye-ring

BLUE-HEADED VIREO

Seasonality

Some bird species in Canada are year-round residents, although a few individuals of these species move away from where they hatched at some time in the year. However, a large number of Canadian species are migratory. For example, many songbirds fly from their breeding grounds in Canada's boreal forests to Mexico and northern South America. Some species, such as the American Robin, are partial migrants, meaning that some populations are resident year-round and others migrate out of their breeding range.

NEOTROPICAL MIGRANT
The Blackpoll Warbler breeds in boreal forests, then migrates south to its wintering grounds in the Caribbean, or Central or South America.

Nests and eggs

Most bird species build their own nests while some use old ones; exceptions include the parasitic cowbirds, which lay their eggs in other species' nests. Nest-building is often done by the female alone, but in some species the male may help or even build it himself. Nests range from a simple scrape in the ground with a few added pebbles to an elaborate woven basket-like structure. Plant matter forms basic nest material.

EGG CUP
Robins lay their eggs in a cup lined with grass stems, built either in shrubs or trees.

NATURAL CAVITY
This Northern Saw-whet Owl is nesting in a tree cavity, likely excavated by a woodpecker.

NEST BOX
Cavity-nesting bluebirds will nest in human made structures.

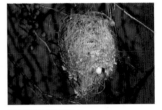

COMPLEX WEAVE
Orioles weave intricate nests from dried plant material and hang them high up in trees.

Egg shapes

There are six basic egg shapes among birds. The most common egg shapes are longitudinal or elliptical. Many cliff-nesting birds lay pear-shaped eggs; if an egg rolls, it does so in a tight circle and remains on the ledge. Spherical eggs with irregular red blotches are characteristic of birds of prey. Pigeons and doves lay white oval eggs, and the eggs of many songbirds are conical and have a variety of dark markings on a pale background. Owls lay the roundest eggs.

PEAR SHAPED

LONGITUDINAL

ELLIPTICAL

OVAL

CONICAL

SPHERICAL

COLOR AND SHAPE
Birds' eggs vary widely in terms of shape, colors, and markings. The American Robin's egg shown above is a beautiful blue.

Gamebirds

This diverse and adaptable group of birds spends most of the time on the ground, springing loudly into the air when alarmed. Among the most terrestrial of all gamebirds, quails are also renowned for their great sociability, often forming large family groups, or "coveys," of up to 100 birds. Grouse are the most numerous and widespread of the gamebirds, and often possess patterns that match their surroundings, providing camouflage from enemies both animal and human. Native to Eurasia, pheasants and partridges were introduced into North America in the 19th and 20th centuries to provide additional targets for recreational hunters. Some adapted well and now thrive in established populations.

SNOW BIRD
The Rock Ptarmigan's winter plumage camouflages it against the snow, helping to hide it from predators.

California Quail

Cellipepla californica

The California Quail thrives in a wide variety of habitats and is increasingly common in parks and suburban habitats. This adaptability, and its popularity among hunters, has led to the California Quail being introduced throughout southern British Columbia and the western US, as well as other areas outside North America.

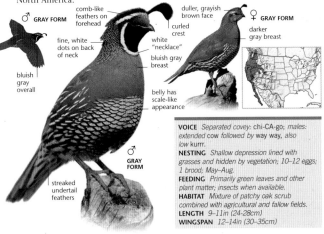

VOICE *Separated covey:* chi-CA-go; *males: extended* cow *followed by* way way, *also low* kurrr.
NESTING *Shallow depression lined with grasses and hidden by vegetation; 10–12 eggs; 1 brood; May–Aug.*
FEEDING *Primarily green leaves and other plant matter; insects when available.*
HABITAT *Mixture of patchy oak scrub combined with agricultural and fallow fields.*
LENGTH *9–11in (24–28cm)*
WINGSPAN *12–14in (30–35cm)*

Northern Bobwhite

Colinus virginianus

This small, plump, chicken-like bird is widely distributed across southwestern Ontario and the eastern US states. When flushed from groundcover, it erupts in "coveys" of 10 to 20 birds and disperses in many directions. Birds raised in captivity are released to supplement wild populations for hunting.

VOICE *Breeding males: whistled* bob-WHITE *or* bob-bob-WHITE; *call to reunite flock:* hoi-lee *and* hoi.
NESTING *Depression on ground lined with plant matter; 10–15 eggs; sometimes multiple broods per season; Jan–Mar.*
FEEDING *Seeds, buds, leaves, and insects, snails, and spiders, when available.*
HABITAT *Agricultural fields; mixed young forests, fields, and brushy hedges.*
LENGTH *8–10in (20–25cm)*
WINGSPAN *11–14in (28–35cm)*

Wild Turkey

Meleagris gallopavo

The largest gamebird in North America, the Wild Turkey was eliminated from most of its original range by the early 1900s due to over-hunting and habitat destruction. Since then, habitat restoration and the subsequent reintroduction of Wild Turkeys has been very successful.

iridescent bronze-and-purplish body

dark overall

♀

black-and-white barred wings

rusty tail with black band

♂ EAST

tail fanned in display

unfeathered blue-and-red head

large red wattles

hair-like "beard" on breast

♂ WEST

dark body, with bronze iridescence

VOICE Males: gobble, *especially during courtship;* females: yelps, clucks, and purrs.
NESTING *Scrape on ground lined with grass, near protective cover; 10–15 eggs; 1 brood; Mar–Jun.*
FEEDING *Omnivorous; acorns, vegetation, and insects from forest floor or agricultural fields.*
HABITAT *Mixed mature woodlands, agricultural fields; grasslands, close to swamps.*
LENGTH 2¾–4ft (0.9–1.2m)
WINGSPAN 4–5ft (1.2–1.5m)

Spruce Grouse

Falcipennis canadensis

The Spruce Grouse's lack of wariness when approached has earned it the name "fool hen." Its diet of pine needles causes the intestinal tract to expand in order to accommodate a larger volume of food to compensate for its low nutritional value. There are two different subspecies of Spruce Grouse (*F. c. canadensis* and *F. c. franklinii*), both of which have red and gray forms.

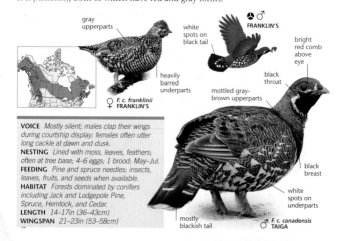

gray upperparts

white spots on black tail

♂♂ FRANKLIN'S

bright red comb above eye

heavily barred underparts

♀ *F. c. franklinii* FRANKLIN'S

mottled gray-brown upperparts

black throat

black breast

white spots on underparts

mostly blackish tail

♂ *F. c. canadensis* TAIGA

VOICE *Mostly silent; males clap their wings during courtship display; females often utter long cackle at dawn and dusk.*
NESTING *Lined with moss, leaves, feathers; often at tree base; 4–6 eggs; 1 brood; May–Jul.*
FEEDING *Pine and spruce needles; insects, leaves, fruits, and seeds when available.*
HABITAT *Forests dominated by conifers including Jack and Lodgepole Pine, Spruce, Hemlock, and Cedar.*
LENGTH 14–17in (36–43cm)
WINGSPAN 21–23in (53–58cm)

Ruffed Grouse

Bonasa umbellus

The Ruffed Grouse is perhaps the most widespread gamebird in North America. There are two color forms, rufous and gray, both allowing the birds to remain camouflaged and undetected on the forest floor, until they eventually burst into the air in an explosion of whirring wings. The male is well known for his extraordinary wing beating or "drumming" display, which he performs year-round, but most frequently in the spring.

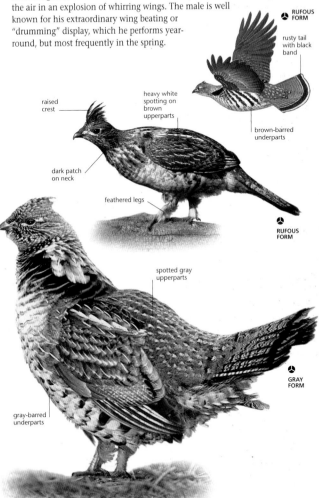

RUFOUS FORM

rusty tail with black band

brown-barred underparts

heavy white spotting on brown upperparts

raised crest

dark patch on neck

feathered legs

RUFOUS FORM

spotted gray upperparts

GRAY FORM

gray-barred underparts

VOICE *Hissing notes, and soft purrt, purrt, purrt when alarmed; males "drumming" display when heard from a distance resembles small engine starting, thump... thump... thump... thuthuthuth.*
NESTING *Shallow, leaf-lined bowl set against tree or fallen log in forest; 6–14 eggs; 1 brood; Mar–Jun.*
FEEDING *Leaves, buds, and fruit from the ground; occasionally insects.*
HABITAT *Young, mixed habitat forests.*
LENGTH *17–20in (43–51cm)*
WINGSPAN *20–23in (51–58cm)*

Greater Sage Grouse

Centrocercus urophasianus

Each spring, male Greater Sage Grouse gather on communal sites, known as leks, where they compete for females with spectacular courtship displays. A lek may contain as many as 40 males. By far the largest native North American grouse, its populations have declined as human encroachment on sagebrush habitats has increased.

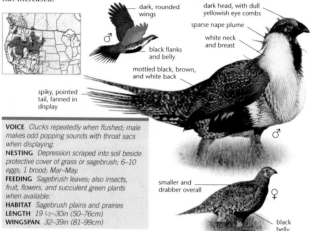

dark, rounded wings

dark head, with dull yellowish eye combs

sparse nape plume

white neck and breast

black flanks and belly

mottled black, brown, and white back

♂

spiky, pointed tail, fanned in display

VOICE *Clucks repeatedly when flushed; male makes odd popping sounds with throat sacs when displaying.*
NESTING *Depression scraped into soil beside protective cover of grass or sagebrush; 6–10 eggs; 1 brood; Mar–May.*
FEEDING *Sagebrush leaves; also insects, fruit, flowers, and succulent green plants when available.*
HABITAT *Sagebrush plains and prairies*
LENGTH 19 ½–30in (50–76cm)
WINGSPAN 32–39in (81–99cm)

smaller and drabber overall

♀

black belly

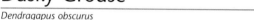

Dusky Grouse Ⓢ

Dendragapus obscurus

Once considered a Blue Grouse subspecies, the Dusky Grouse was reclassified as a species in its own right. Males have reddish or purple air sacs, and their courtship displays are primarily ground-based and quieter than those of the Sooty Grouse. The Dusky Grouse also has a plainer tail, lacking the grayer tip of the Sooty.

pale underwing

barred crown and neck

small bill

♀

mottled brown back

♂

broad, rounded black tail

red wattle over eye

bare red or purple air sacs

gray belly

♂ DISPLAY

short, plain brown wings

gray underparts

white scales on flanks

VOICE *Five soft hoots; also a hiss, growl, and cluck; females: a whinnying cry.*
NESTING *Shallow scrape, lined with dead grass, leaves, or other plants, under shrubs or against rocks or logs; 7–10 eggs; 1 brood; Mar–May.*
FEEDING *Leaves, flowers, fruit, also some insects; evergreen needles, buds, and cones.*
HABITAT *High or mid-altitude open forests and shrublands.*
LENGTH 16–20in (41–51cm)
WINGSPAN 25–28in (64–71cm)

Sooty Grouse (S)

Dendragapus fuliginosus

The Sooty Grouse, like the Dusky Grouse, was reclassified as a separate species. Restricted to coastal mountain ranges, the male Sooty Grouse also differs from the Dusky Grouse in its yellow air sacs and its courtship displays often performed in trees. Females and chicks have a browner overall appearance to their plumage.

gray band at tip of tail

♂

heavily mottled

♀

dark cheek patch above pale throat

deep-red wattle

yellow air sacs on side of neck

barred tail with gray tip

dark upperparts

♂

short, stiffly curved wings

dark underparts

VOICE *Loud six-syllable hooting; also growl, hiss, cluck, purr.*
NESTING *Shallow depression lined with dead vegetation, usually under small pine trees; 5–8 eggs; 1 brood; Mar–May.*
FEEDING *Evergreen needles, especially Douglas Fir; also leaves, grasses, fruit, insects.*
HABITAT *Open areas with grassland, forest clearings, and shrubs for breeding; thick evergreen forests in winter.*
LENGTH *16–20in (41–51cm)*
WINGSPAN *25–28in (64–71cm)*

Sharp-tailed Grouse (S)

Tympanuchus phasianellus

The Sharp-tailed Grouse is able to adapt to a great variety of habitats, moving between grassland summer habitats and woodland winter habitats. Elements of this grouse's spectacular courtship display have been incorporated into the culture and dance of Native American people, including foot stomping and tail feather rattling. The Sharp-tailed Grouse is the provincial bird of Saskatchewan.

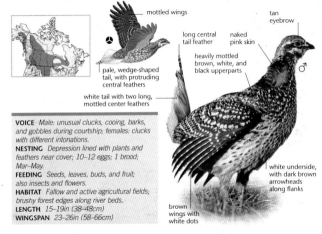

mottled wings

tan eyebrow

long central tail feather

naked pink skin

heavily mottled brown, white, and black upperparts

♂

pale, wedge-shaped tail, with protruding central feathers

white tail with two long, mottled center feathers

white underside, with dark brown arrowheads along flanks

brown wings with white dots

VOICE *Male: unusual clucks, cooing, barks, and gobbles during courtship; females: clucks with different intonations.*
NESTING *Depression lined with plants and feathers near cover; 10–12 eggs; 1 brood; Mar–May.*
FEEDING *Seeds, leaves, buds, and fruit; also insects and flowers.*
HABITAT *Fallow and active agricultural fields; brushy forest edges along river beds.*
LENGTH *15–19in (38–48cm)*
WINGSPAN *23–26in (58–66cm)*

White-tailed Ptarmigan ⓢ

Lagopus leucura

The White-tailed Ptarmigan is the smallest and most southerly of the three
North American ptarmigans. In the winter, its almost completely white
plumage—unique among the gamebird species—blends it in perfectly to its icy,
mountainous home. The feathers on its feet help to prevent the bird from
sinking into the snow.

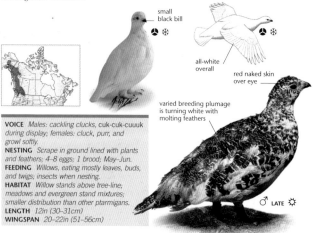

small
black bill

all-white
overall

red naked skin
over eye

varied breeding plumage
is turning white with
molting feathers

♂ LATE ☼

VOICE *Males: cackling clucks, cuk-cuk-cuuuk
during display; females: cluck, purr, and
growl softly.*
NESTING *Scrape in ground lined with plants
and feathers; 4–8 eggs; 1 brood; May–Jun.*
FEEDING *Willows, eating mostly leaves, buds,
and twigs; insects when nesting.*
HABITAT *Willow stands above tree-line;
meadows and evergreen stand mixtures;
smaller distribution than other ptarmigans.*
LENGTH *12in (30–31cm)*
WINGSPAN *20–22in (51–56cm)*

Rock Ptarmigan ⓢ

Lagopus muta

The Rock Ptarmigan is well known for its distinctive seasonal variation in
plumage, which helps to camouflage it. Although some birds make a short
migration to more southern wintering grounds, many remain on their breeding
grounds year-round. The Rock Ptarmigan is the official bird of Nunavut Territory.

brown-and-
black barring

black line between
eye and bill

mostly gray
upperparts

♀ ☼

mottled
belly

♂ ☼

white
wings

gray
wing
patch

red
comb

♂ ❄

small,
round
head

small,
delicate
bill

"salt-and-pepper"
barring on gray
upperparts

♂ ☼

white
belly

feathered
feet

VOICE *Quiet; male: raspy krrrh, also growls
and clucks.*
NESTING *Small scrape or natural depression,
lined with plant matter, often exposed; 8–10
eggs; 1 brood; Apr–Jun.*
FEEDING *Buds, seeds, flowers, and leaves,
especially birch and willow; insects.*
HABITAT *Dry, rocky tundra and shrubby
ridge tops; open meadow edges and dense
evergreen stands along waterways in winter.*
LENGTH *12½–15½in (32–40cm)*
WINGSPAN *19½–23½in (50–60cm)*

Willow Ptarmigan

Lagopus lagopus

The most common of the three ptarmigan species, the Willow Ptarmigan also undertakes the longest migration of the group from its wintering to its breeding range. The Willow Ptarmigan is an unusual gamebird species, as male and female remain bonded throughout the chick-rearing process, in which the male is an active participant.

black bill

♂ ❄

reddish brown body

black bill

red comb

♂ ☼

all-white body

rich reddish brown body

dark, scaly bars

white belly

feathered feet

VOICE *Purrs, clucks, hissing, meowing noises; Kow-Kow-Kow call given before flushing.*
NESTING *Shallow bowl scraped in soil, lined with plant matter, with overhead cover; 8–10 eggs; 1 brood; Mar–May.*
FEEDING *Mostly buds, stems, and seeds; also flowers, insects, and leaves.*
HABITAT *Arctic, sub-Arctic, and sub-alpine tundra; willow thickets along river corridors; low woodlands.*
LENGTH *14–17½in (35–44cm)*
WINGSPAN *22–24in (56–61cm)*

Chukar ⓢ

Alectoris chukar

A native of Eurasia, the Chukar was introduced as a gamebird during the mid-20th century in six Canadian provinces and more than 40 US states. The species succeeded in limited areas, especially on steep mountain slopes in the West. Chukars form large communal groups, or crèches, of up to 100 young birds, with 10–12 adults overseeing them.

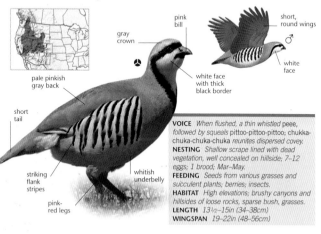

pink bill

gray crown

short, round wings

pale pinkish gray back

♂

white face

white face with thick black border

short tail

striking flank stripes

whitish underbelly

pink-red legs

VOICE *When flushed, a thin whistled* peee, *followed by squeals* pittoo-pittoo-pittoo; chukka-chuka-chuka-chuka *reunites dispersed covey.*
NESTING *Shallow scrape lined with dead vegetation, well concealed on hillside; 7–12 eggs; 1 brood; Mar–May.*
FEEDING *Seeds from various grasses and succulent plants; berries; insects.*
HABITAT *High elevations; brushy canyons and hillsides of loose rocks, sparse bush, grasses.*
LENGTH 13½–15in (34–38cm)
WINGSPAN 19–22in (48–56cm)

Gray Partridge ⓢ

Perdix perdix

This Eurasian native was introduced to North America in the late 18th century. Benefiting from the mixture of agricultural and fallow fields that resulted from long-term conservation programs, its population is stable or expanding in the west. The isolated eastern populations are declining due to changes in land use.

brown, rounded wings

cinnamon face

rusty head

gray neck and chest with fine black barring

dark cinnamon tail

gray back with fine barring

horseshoe-shaped belly patch

underparts gray overall

chestnut barred gray flanks

VOICE *Short, quick* kuk-kuk-kuk *when alarmed;* prruk-prruk *between adults and young when threatened.*
NESTING *Shallow depression lined with vegetation, usually in hedgerows; 14–18 eggs; 1 brood; Mar–May.*
FEEDING *Seeds and row crops (corn and wheat); leaves in spring; insects when breeding.*
HABITAT *Agricultural fields including corn, wheat, and oats; hedgerows, fallow grasslands.*
LENGTH 11–13in (28–33cm)
WINGSPAN 17–20in (43–51cm)

Ring-necked Pheasant

S

Phasianus colchicus

A native of Asia, the Ring-necked Pheasant was originally introduced in North America for recreational hunting purposes, and is now widely distributed across North America. Birds released after being bred in captivity are used to supplement numbers for hunting purposes. In the wild, several females may lay eggs in the same nest—a phenomenon called "egg-dumping." There is a less common dark form, which can be distinguished principally because it lacks the distinctive white band around the neck.

short, round wings

long tail

♂

pale rump

pointed tail

♀

bold black markings

pale brown body

♀

iridescent ear tufts

red face wattles

white neck ring

iridescent bronze sheen

long, pointed tail

barred underparts

♂

VOICE *Male: loud, raucous, explosive double note, Karrk-KORK; both sexes cackle when flushed.*
NESTING *Shallow bowl composed of grasses, usually in tall grass or among low shrubs; 7–15 eggs; 1 brood; Mar–Jun.*
FEEDING *Corn and other grain, seeds, fruit, row crops, grass, leaves and shoots; insects.*
HABITAT *Agricultural fields, particularly corn, fallow fields, and hedgerows; cattail marshes and wooded river bottoms.*
LENGTH *19½–28in (50–70cm)*
WINGSPAN *30–34in (76–86cm)*

Waterfowl

Geese are more terrestrial than either swans or ducks, often being seen grazing on dry land. Like swans, geese pair for life. They are also highly social, and most species are migratory, flying south for the winter in large flocks.

Swans are essentially large, long-necked geese. Ungainly on land, they are extremely graceful on water. When feeding, a swan stretches its long neck to reach water plants at the bottom, submerging up to half its body as it does so.

Ducks are more varied than swans or geese, and are loosely grouped by their feeding habits. Dabblers, or puddle ducks, eat plants and other edible matter like snails by upending on the surface of shallow water. Diving ducks, by contrast, dive deep underwater for their food.

GAGGLING GEESE
Gregarious Snow Geese form large, noisy flocks during migration and on winter feeding grounds.

Greater White-fronted Goose Ⓢ

Anser albifrons

The widespread Greater White-fronted Goose is easily distinguished by its black-barred belly and the patch of white at the base of its bill. The "Tundra" (*A. a. frontalis*), makes up the largest population, breeding across northwestern Canada and western Alaska. The "Tule" (*A. a. gambeli*) occurs in the fewest numbers and is restricted in range to northwest Canada.

gray wing feathers

darker chocolate-brown upperparts

white rump band

larger body

white tip to tail

pink bill with white base

longer legs, bill, and neck

white flank streak

brownish gray head

A.a gambeli TULE

♂

A.a frontalis TUNDRA

bright orange legs

brown underparts with black bands

VOICE *Laugh-like* klow-yo *or* klew-yo-yo; *very musical in a flock.*
NESTING *Bowl-shaped nest made of plant material, lined with down, near water; 3–7 eggs; 1 brood; May–Aug.*
FEEDING *Sedges, grasses, berries, and plants on both land and water in summer; grasses, seeds, and grains in winter.*
HABITAT *Tundra ponds and lakes, rocky fields, and grassy slopes for nesting.*
LENGTH *25–32in (64–81cm)*
WINGSPAN *4¼–5¼ft (1.3–1.6m)*

Snow Goose Ⓢ

Chen caerulescens

The abundant Snow Goose has two subspecies. The "Greater" (*A. c. atlantica*) is slightly larger and breeds further east. The smaller "Lesser" (*A. c. caerulescens*) breeds further west. Snow Geese have two color forms—white and "blue" (actually dark grayish brown with a white head), and there are also intermediate forms. Heads are often stained rusty-brown from minerals in soil.

gray wing patch

blackish brown back

elongated, white head

dark belly

black patch on long bill

WHITE

pale wing feathers

long neck

BLUE FORM

white upperparts

VOICE *Nasal* whouk, kowk, *or* kow-luk, *higher-pitched* heenk; *feeding call* hu-hu-hur.
NESTING *Scrapes on hummock, lined with plant material and down; 2–6 eggs; 1 brood; May–Jul.*
FEEDING *Aquatic and terrestrial vegetation, including stems, seeds, leaves, tubers, and roots; grain.*
HABITAT *Tundra while breeding and interior valleys and coastal marshes in winter.*
LENGTH *27–33in (69–83cm)*
WINGSPAN *4¼–5½ft (1.3–1.7m)*

pink legs

WHITE FORM

Ross's Goose

Chen rossii

Ross's Goose is not much bigger than a Mallard; like the Snow Goose, it also has a rare "blue" form. About 95 percent of Ross's Geese nest at a single sanctuary in Arctic Canada; the rest breed along Hudson Bay and at several island locations. Hunting drastically reduced the population in the early 1950s, but the species has rebounded.

black wing tips

WHITE

round head

short, triangular bill

BLUE FORM

mostly dark brown upperparts

short, deeply furrowed neck

clean white upperparts

white rump and tail

WHITE FORM

reddish pink legs

VOICE *Keek keek keeek, higher-pitched than Snow Goose; also a harsh, low kork or kowk; quiet while feeding.*
NESTING *Plant materials placed on ground, usually in colonies with Lesser Snow Geese; 3–5 eggs; 1 brood; Jun–Aug.*
FEEDING *Grasses, sedges, and small grains.*
HABITAT *Tundra while breeding; agricultural fields and grasslands in winter; roosts overnight in wetlands.*
LENGTH *22½–25in (57–64cm)*
WINGSPAN *3¼ft (1.1m)*

Cackling Goose

Branta hutchinsii

The Cackling Goose is distinguished from the Canada Goose by its smaller size, its short stubby bill, steep forehead, and short neck. There are four subspecies of Cackling Goose, which vary in breast color, ranging from dark on *B. h. minima*, fairly dark on *B. h. leucopareia*, and pale on *B. h. hutchinsii*.

plain grayish brown wings

white chin strap

small stubby bill

white u-shaped patch on rump

small, black head

B. h. hutchinsii

no black under chin

black tail

pale breast

dark brown breast

B. h. minima

VOICE *Males: honk or bark; females: higher pitched hrink; also high-pitched yelps.*
NESTING *Scrape lined with available plant matter and down; 2–8 eggs; 1 brood; May–Aug.*
FEEDING *Plants in summer; in winter, grass in livestock and dairy pastures, and agricultural fields.*
HABITAT *Rocky tundra slopes while breeding; pastures and agricultural fields in winter.*
LENGTH *21½–30in (55–75cm);*
WINGSPAN *4¼–5ft (1.3–1.5m)*

Canada Goose

Branta canadensis

The Canada Goose is the most common, widespread, and familiar goose in North America. Given its colossal range, it is not surprising that the Canada Goose has much geographic variation, and 12 subspecies have been recognized. With the exception of the Cackling Goose, from which it has recently been separated, it is difficult to confuse it, with its distinctive white chin strap, black head and neck, and grayish brown body, with any other species of goose. It is a monogamous species, and once pairs are formed, they usually stay together for life.

plain grayish brown wings with darker flight feathers

white u-shaped patch on rump

smaller, white chin strap

dark brown overall

black head

broad white chin strap

very long neck

grayish brown upperparts and sides

paler upper breast

white undertail feathers

VOICE *Males:* honk or bark; *females:* higher pitched hrink.
NESTING *Scrape lined with plant matter and down, near water; 1–2 broods; 2–12 eggs; May–Aug.*
FEEDING *Grasses, sedges, leaves, seeds, agricultural crops and berries; insects.*
HABITAT *Inland near water, including grassy urban areas, marshes, prairies, parkland, forests, and tundra; agricultural fields, saltwater marshes, lakes, and rivers in winter.*
LENGTH 2¼–3½ft (0.7–1.1m)
WINGSPAN 4¼–5½ft (1.3–1.7m)

Brant

Ⓢ

Branta bernicla

A small-billed, dark, stocky sea goose, the Brant winters on the east and west coasts of North America: the pale-bellied "Atlantic" Brant (*B. b. hrota*) in the east, and the darker "Black" Brant (*B. b. nigricans*) in the west. An intermediate gray-bellied form, not yet formally named, breeds in the Canadian archipelago and winters in Boundary Bay, British Columbia

pale bars across wings

small, white "necklace" not crossing throat

black neck stops abruptly at breast

barred flanks with pale belly

broad white necklace crosses throat

white rump

black neck and head

EASTERN

B. b. hrota **EASTERN**

dark gray-brown upperparts

grayish white flank patch

black chest

bold, barred flanks

bold, white rump

very dark belly

B. b. nigricans **WESTERN**

VOICE Nasal, harsh cruk; series of cut cut cut cronk, ending in an upward inflection.
NESTING Scrape lined with plant matter and down on islands or gravel spits; 3–5 eggs; 1 brood; May–Jul.
FEEDING Grass and sedges when nesting; eelgrass in winter; green algae, salt marsh plants, and mollusks.
HABITAT High Arctic tundra while breeding; Pacific and Atlantic coasts in winter.
LENGTH 22–26in (56–66cm)
WINGSPAN 3½–4ft (1.1–1.2m)

Mute Swan

Ⓢ

Cygnus olor

The Mute Swan was introduced from Europe due to its graceful appearance on water, if not on land, and easy domestication. However, this is an extremely territorial and aggressive bird. When threatened, it points its bill downwards, arches its wings, hisses, and then attacks. Displacement of native waterfowl species and overgrazing have led to efforts to reduce its numbers in North America.

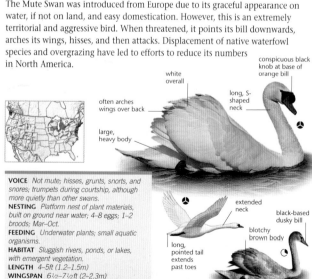

conspicuous black knob at base of orange bill

white overall

long, S-shaped neck

often arches wings over back

large, heavy body

extended neck

black-based dusky bill

blotchy brown body

long, pointed tail extends past toes

VOICE Not mute; hisses, grunts, snorts, and snores; trumpets during courtship, although more quietly than other swans.
NESTING Platform nest of plant materials, built on ground near water; 4–8 eggs; 1–2 broods; Mar–Oct.
FEEDING Underwater plants; small aquatic organisms.
HABITAT Sluggish rivers, ponds, or lakes, with emergent vegetation.
LENGTH 4–5ft (1.2–1.5m)
WINGSPAN 6½–7½ft (2–2.3m)

Trumpeter Swan (S)

Cygnus buccinator

North America's quintessential swan and heaviest waterfowl, the Trumpeter has made a remarkable comeback after numbers were severely reduced by hunting; by the mid-1930s, fewer than a hundred were known to exist. Reintroduction efforts were made in Ontario and the upper Midwest in the US to re-establish the species. The Trumpeter Swan's characteristic call is usually the best way to identify it.

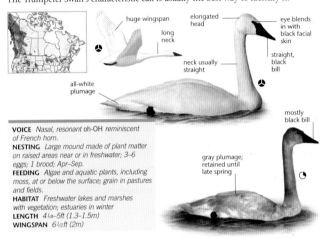

huge wingspan

long neck

elongated head

eye blends in with black facial skin

straight, black bill

neck usually straight

all-white plumage

mostly black bill

gray plumage; retained until late spring

VOICE Nasal, resonant oh-OH reminiscent of French horn.
NESTING Large mound made of plant matter on raised areas near or in freshwater; 3–6 eggs; 1 brood; Apr–Sep.
FEEDING Algae and aquatic plants, including moss, at or below the surface; grain in pastures and fields.
HABITAT Freshwater lakes and marshes with vegetation; estuaries in winter
LENGTH 4¼–5ft (1.3–1.5m)
WINGSPAN 6½ft (2m)

Tundra Swan (S)

Cygnus columbianus

Nesting in the Arctic tundra, this well-named species is North America's most widespread and smallest swan. Two populations exist, with one wintering in the West, and the other along the East Coast. The Tundra Swan can be confused with the Trumpeter Swan, but their different calls immediately distinguish the two species.

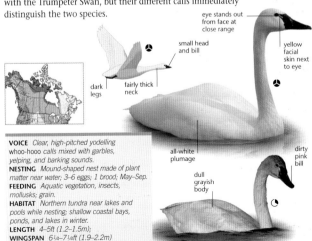

eye stands out from face at close range

yellow facial skin next to eye

small head and bill

dark legs

fairly thick neck

all-white plumage

dull grayish body

dirty pink bill

VOICE Clear, high-pitched yodelling whoo-hooo calls mixed with garbles, yelping, and barking sounds.
NESTING Mound-shaped nest made of plant matter near water; 3–6 eggs; 1 brood; May–Sep.
FEEDING Aquatic vegetation, insects, mollusks; grain.
HABITAT Northern tundra near lakes and pools while nesting; shallow coastal bays, ponds, and lakes in winter.
LENGTH 4–5ft (1.2–1.5m);
WINGSPAN 6¼–7¼ft (1.9–2.2m)

Wood Duck ⓢ

Aix sponsa

The male Wood Duck is unmistakable, with its gaudy plumage, red eye and bill, and its helmet-shaped profile. The Wood Duck is typically found on swamps, shallow lakes, ponds, and park settings that are surrounded by trees. When swimming, it jerks its head front to back. Of all waterfowl, this is the only species that regularly raises two broods each season.

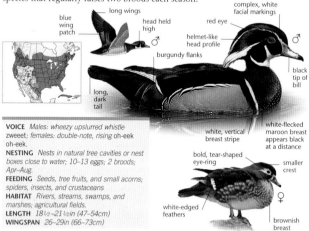

blue wing patch

long wings

head held high

complex, white facial markings

red eye ♂

helmet-like head profile

burgundy flanks

black tip of bill

long, dark tail

white, vertical breast stripe

white-flecked maroon breast appears black at a distance

bold, tear-shaped eye-ring

smaller crest

white-edged feathers

brownish breast ♀

VOICE *Males: wheezy upslurred whistle zweeet; females: double-note, rising oh-eek oh-eek.*
NESTING *Nests in natural tree cavities or nest boxes close to water; 10–13 eggs; 2 broods; Apr–Aug.*
FEEDING *Seeds, tree fruits, and small acorns; spiders, insects, and crustaceans*
HABITAT *Rivers, streams, swamps, and marshes; agricultural fields.*
LENGTH 18½–21½in (47–54cm)
WINGSPAN 26–29in (66–73cm)

Gadwall ⓢ

Anas strepera

Despite being common and widespread, Gadwalls are often overlooked because of their retiring behavior and relatively quiet vocalizations. This dabbling duck is slightly smaller and more delicate than the Mallard, yet female Gadwalls are often mistaken for female Mallards. Gadwalls associate with other species, especially in winter.

conspicuous white patch ♂❄

mostly white underwings

brown, rounded head

black bill

dark grayish overall ♂❄

white belly

black uppertail

orange-yellow legs

finely patterned gray flanks and breast

dark eyestripe

white wing patch

brown, scalloped back ♀

VOICE *Low, raspy meep or reb in quick succession; females: high-pitched, nasal quack; high-pitched peep, or pe-peep; tickety-tickety-tickety chatter while feeding.*
NESTING *Bowl nest made of plant material in a scrape; 8–12 eggs; 1 brood; Apr–Aug.*
FEEDING *Seeds, aquatic vegetation, and invertebrates, including mollusks and insects.*
HABITAT *Shallow wetlands, reservoirs, ponds, lakes, and rivers.*
LENGTH 18–22½ in (46–57cm)
WINGSPAN 33in (84cm)

American Wigeon

Anas americana

Often found in mixed flocks with other ducks, the American Wigeon is a common and widespread, medium-sized dabbling duck. This bird is an opportunist that loiters around other diving ducks and coots, feeding on the vegetation they dislodge. It is more social during migration and in the nonbreeding season than when breeding. The male's cream-colored forehead is distinctive.

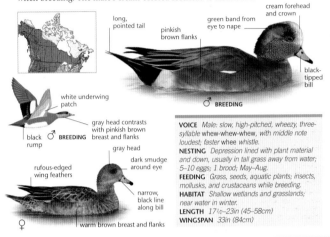

long, pointed tail

pinkish brown flanks

green band from eye to nape

cream forehead and crown

black-tipped bill

♂ **BREEDING**

white underwing patch

black ♂ **BREEDING** rump

gray head contrasts with pinkish brown breast and flanks

rufous-edged wing feathers

gray head

dark smudge around eye

narrow, black line along bill

♀

warm brown breast and flanks

VOICE *Male: slow, high-pitched, wheezy, three-syllable* whew-whew-whew, *with middle note loudest; faster* whee *whistle.*
NESTING *Depression lined with plant material and down, usually in tall grass away from water; 5–10 eggs; 1 brood; May–Aug.*
FEEDING *Grass, seeds, aquatic plants; insects, mollusks, and crustaceans while breeding.*
HABITAT *Shallow wetlands and grasslands; near water in winter.*
LENGTH *17½–23in (45–58cm)*
WINGSPAN *33in (84cm)*

American Black Duck

Anas rubripes

Closely related to the Mallard, the American Black Duck prefers forested locations in contrast to the Mallard's more open habitats. Over the years, as the East was deforested and trees were planted in the Midwest, the two species' habitats have become less distinct, resulting in interbreeding. When breeding, American Black Duck males can be seen chasing away other males.

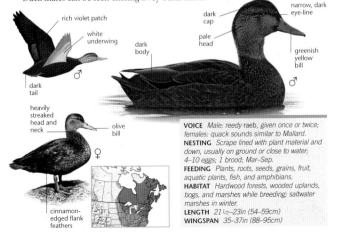

rich violet patch

white underwing

dark body

♂

dark tail

narrow, dark eye-line

dark cap

pale head

greenish yellow bill

♂

heavily streaked head and neck

olive bill

♀

cinnamon-edged flank feathers

VOICE *Male: reedy* raeb, *given once or twice; females: quack sounds similar to Mallard.*
NESTING *Scrape lined with plant material and down, usually on ground or close to water; 4–10 eggs; 1 brood; Mar–Sep.*
FEEDING *Plants, roots, seeds, grains, fruit, aquatic plants, fish, and amphibians.*
HABITAT *Hardwood forests, wooded uplands, bogs, and marshes while breeding; saltwater marshes in winter.*
LENGTH *21½–23in (54–59cm)*
WINGSPAN *35–37in (88–95cm)*

Mallard

Ⓢ

Anas platyrhynchos

The Mallard is perhaps the most familiar of all ducks, and occurs in the wild all across the Northern Hemisphere. It is the ancestor of most domestic ducks, and hybrids between the wild and domestic forms are frequently seen in city lakes and ponds, often with patches of white on the breast. Mating is generally a violent affair, but outside the breeding season the wild species is strongly migratory and gregarious, sometimes forming large flocks that may join with other species. In some cities they even nest in backyards with swimming pools.

broad-based wings

♂ ❅

short, round, pale tail

heavy body

♀

brown underparts

whitish outer tail feathers

orange bill with blackish patch

dark eye-line and cap

♀

yellowish brown back

mottled brown belly

bright yellow bill

metallic green head

short, black curls above white tail

warm gray body

blue wing patch

narrow, white neck collar

♂ ❅

chestnut-brown breast

VOICE Male: quiet raspy *raab*; during courtship a high-pitched whistle; females: *quack*.
NESTING Scrape lined with plant matter, usually near water or on floating vegetation; 6–15 eggs; 1 brood; Feb–Sep.
FEEDING Insects, crustaceans, mollusks, and earthworms when breeding; seeds, acorns, agricultural crops, and aquatic vegetation.
HABITAT Shallow water in marshes, ponds, and ditches; city parks and reservoirs.
LENGTH 19½–26in (50–65cm)
WINGSPAN 32–37in (82–95cm)

Blue-winged Teal (S)

Anas discors

With a bold white crescent between bill and eye on its otherwise slate-gray head and neck, the male Blue-winged Teal is quite distinctive. The Blue-winged and Cinnamon Teals, along with the Northern Shoveler, constitute the three "blue-winged" ducks; a feature that is conspicuous during flight. The Cinnamon and the Blue-winged Teals are almost identical genetically and sometimes interbreed.

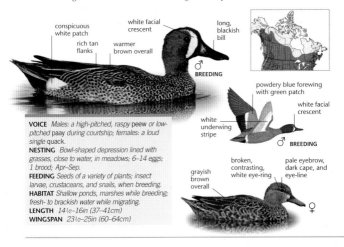

conspicuous white patch

white facial crescent

rich tan flanks

warmer brown overall

long, blackish bill

♂ BREEDING

powdery blue forewing with green patch

white facial crescent

white underwing stripe

♂ BREEDING

grayish brown overall

broken, contrasting, white eye-ring

pale eyebrow, dark cape, and eye-line

♀

VOICE *Males: a high-pitched, raspy peew or low-pitched paay during courtship; females: a loud single quack.*
NESTING *Bowl-shaped depression lined with grasses, close to water, in meadows; 6–14 eggs; 1 brood; Apr–Sep.*
FEEDING *Seeds of a variety of plants; insect larvae, crustaceans, and snails, when breeding.*
HABITAT *Shallow ponds, marshes while breeding; fresh- to brackish water while migrating.*
LENGTH *14½–16in (37–41cm)*
WINGSPAN *23½–25in (60–64cm)*

Cinnamon Teal (S)

Anas cyanoptera

The male Cinnamon Teal is unmistakable in its overall rusty brown color and blazing red eyes. This fairly small, dabbling duck is common in southwestern Canada and western US, and even seen in tiny roadside pools. Closely related to both the Northern Shoveler and Blue-winged Teal, the Cinnamon Teal's wing pattern is indistinguishable from that of the latter.

solid cinnamon color

conspicuous orange to red eye

♂

powdery blue forewing

♂

white underwing stripe

long, spoon-shaped, black bill

warm brown, upperparts with rust tinge

plain face pattern

shoveler-like bill

♀

dull yellow legs

VOICE *Males: snuffled chuk chuk chuk; females: loud quack, soft gack gack gack ga.*
NESTING *Shallow depression lined with grass near water; 4–16 eggs; 1 brood; Mar–Sep.*
FEEDING *Seeds of many plant species; aquatic insects, crustaceans, and snails, when breeding.*
HABITAT *Freshwater and brackish marshes, reservoirs, ponds, and ditches; tidal estuaries and salt marshes in winter.*
LENGTH *14–17in (36–43cm)*
WINGSPAN *22in (56cm)*

Northern Shoveler

Anas clypeata

The Northern Shoveler is a common, medium-sized, dabbling duck found in North America and Eurasia. It is monogamous—pairs remain together longer than any other dabbler species. Its distinctive long bill is highly specialized to filter food items from the water. Shovelers often form tight feeding groups, swimming close together as they sieve the water for prey.

pale blue wing patch

♂

heavy fronted

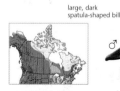

large, dark spatula-shaped bill

yellow eye

♂

dark green head

white breast

chestnut belly and flanks

black-and-white rump

VOICE *Males: nasal, muffled* thuk thuk…thuk thuk; *loud, nasal* paaaay; *females: variety of* quacks, *singly or in series of 4–5 descending notes.*
NESTING *Scrape lined with plant matter and down, in short plants, near water; 6–19 eggs; 1 brood; May–Aug.*
FEEDING *Seeds; small crustaceans, mollusks.*
HABITAT *Wetlands with nearby grasslands; fresh- and saltmarshes and ponds in winter.*
LENGTH *17½–20in (44–51cm)*
WINGSPAN *27–33in (69–84cm)*

dark, narrow eye-line

pale-edged, brown flank feathers

♀

dusky olive-gray to orange bill

brown overall

Northern Pintail

Anas acuta

An elegant, long-necked dabbler, the Northern Pintail has distinctive markings and the longest tail of any freshwater duck. It begins nesting soon after the ice thaws. Northern Pintails were once one of the most abundant prairie breeding ducks, but droughts, combined and habitat reduction on both their wintering and breeding grounds have resulted in a population decline.

green wing patch with buff bar

♂❄

outstretched head and neck

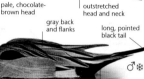

black bill with gray sides

pale, chocolate-brown head

gray back and flanks

long, pointed black tail

♂❄

long neck

white neck and breast

black undertail with white flank patch

VOICE *Males: high-pitched* prrreep prrreep; *lower-pitched wheezy* wheeeee; *females: quiet, harsh quack or* kuk; *loud, repeated* gaak.
NESTING *Scrape lined with plants and down; 3–12 eggs; 1 brood; Apr–Aug.*
FEEDING *Grains, rice, seeds, aquatic weeds, insect larvae, crustaceans, and snails.*
HABITAT *Shallow wetlands or meadows in mountainous forests while breeding; tidal wetlands, saltwater habitats while migrating.*
LENGTH *20–30in (51–76cm)*
WINGSPAN *35in (89cm)*

blackish bill

plain buff face with dark eye

♀

mottled gray-brown body

pointed tail shorter than male

Green-winged Teal

Ⓢ

Anas crecca

The Green-winged Teal, the smallest North American dabbling duck, is slightly smaller than the Blue-winged and Cinnamon Teals, and lacks their blue wing patch. Its population is increasing, apparently because it breeds in more pristine habitats, and farther north, than the prairie ducks. *A. c. carolinensis* males have a conspicuous vertical white bar on their wings.

♂

short neck

gray flanks

green-and-black patch on hindwing

rufous head

dark green ear patch

small, narrow, black bill

black-spotted breast

white vertical bar

♂

yellowish buff undertail feathers

finely detailed pattern

steeper forehead

darker face

♀

VOICE *Males: high-pitched, slightly rolling* crick crick, *similar to a cricket; females: quiet* quack.
NESTING *Shallow scrape on ground lined with vegetation, often in dense vegetation near water; 6–9 eggs; 1 brood; Apr–Sep.*
FEEDING *Seeds, aquatic insects, crustaceans, and mollusks year-round; grain in winter.*
HABITAT *Ponds in woodlands while breeding; inland and coastal marshes, sloughs, and agricultural fields in winter.*
LENGTH *12–15½in (31–39cm)*
WINGSPAN *20½–23in (52–59cm)*

Canvasback

Ⓢ

Aythya valisineria

A large, long-billed diving duck, the Canvasback is a bird of prairie pothole country with a specialized diet of aquatic plants. With legs set toward the rear, it is an accomplished swimmer and diver, and is rarely seen on land. Weather conditions and brood parasitism by Redheads determine how successful the Canvasback's nesting is from year to year.

♂

light gray forewing

high, peaked black crown

rich chestnut head and neck

long neck, held horizontally in flight

belly appears white

black rump and tail

bright red eye

white to pale gray back and flanks

black at both ends

black breast

♂

VOICE *Males: soft cooing noises during courtship; females: grating* krrrrr krrrrrr krrrrr; *both sexes: soft wheezing* rrrr rrrr rrrr.
NESTING *Platform of woven vegetation built over water or occasionally on shore; 8–11 eggs; 1 brood; Apr–Sep.*
FEEDING *Aquatic tubers, buds, root stalks, and shoots, particularly wild celery; snails.*
HABITAT *Shallow wetlands in prairies and tundra; northern forests.*
LENGTH *19–22in (48–56cm)*
WINGSPAN *31–35in (79–89cm)*

extended tear drop

♀

dingy brownish gray upperparts and sides

Redhead (S)

Aythya americana

The Redhead, a medium-sized diving duck, is native only to North America. The male's seemingly gray upperparts and flanks are actually white, with dense, black, wavy markings. The Redhead forages mostly around dusk and dawn, drifting during the day. It lays its eggs in other duck nests more than any other duck species, particularly those of the Canvasback and even other Redheads.

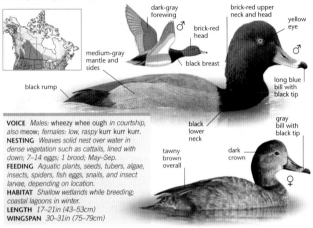

dark-gray forewing ♂

brick-red head

medium-gray mantle and sides

black breast

black rump

brick-red upper neck and head

yellow eye ♂

long blue bill with black tip

gray bill with black tip

black lower neck

dark crown

tawny brown overall

VOICE *Males: wheezy* whee ough *in courtship, also* meow; *females: low, raspy* kurr kurr kurr.
NESTING *Weaves solid nest over water in dense vegetation such as cattails, lined with down; 7–14 eggs; 1 brood; May–Sep.*
FEEDING *Aquatic plants, seeds, tubers, algae, insects, spiders, fish eggs, snails, and insect larvae, depending on location.*
HABITAT *Shallow wetlands while breeding; coastal lagoons in winter.*
LENGTH *17–21in (43–53cm)*
WINGSPAN *30–31in (75–79cm)*

Ring-necked Duck (S)

Aythya collaris

The Ring-necked Duck is a fairly common medium-sized diving duck. The bold white band near its bill tip is easy to see, whereas the thin chestnut ring around the neck can be very difficult to observe. The tall, pointed head is quite distinctive, peaking at the rear of the crown.

tall, peaked head

gray bill with white band at base

yellow eye

thin chestnut ring

rounded gray sides

black neck and breast

dark forewing ♂

bold white underwing

dark brown back

bold white eye-ring

white band on bill

VOICE *Males: normally silent; females: low* kerp kerp.
NESTING *Floating nest built in dense aquatic vegetation, often in marshes; 6–14 eggs; 1 brood; May–Aug.*
FEEDING *Aquatic plant tubers and seeds; aquatic invertebrates such as clams and snails.*
HABITAT *Shallow freshwater marshes and bogs while breeding; swamps, lakes, estuaries, and flooded fields in winter.*
LENGTH *15–18in (38–46cm)*
WINGSPAN *24–25in (62–63cm)*

Greater Scaup

S

Aythya marila

Due to its more restricted range for breeding and wintering, the Greater Scaup is less numerous in North America than the Lesser Scaup. The Greater Scaup forms large, often sexually segregated flocks outside the breeding season. If both scaup species are present, they will also segregate within the flocks according to species. Correct identification is difficult.

blue-gray bill, wider at tip

smooth, round, black head with purple-green gloss

wavy patterned gray back

almost all white sides

♂ BREEDING

gray forewing

♂ NONBREEDING

broad, white wing stripe

bold white patches at base of bill

medium to dark brown overall

♀ NONBREEDING

VOICE *Males: soft, fast, wheezy* week week wheew *during courtship; females: growled monotone* arrrr *series.*
NESTING *Depression lined with grasses and down, with dense cover of vegetation; 6–10 eggs; 1 brood; May–Sep.*
FEEDING *Dives for aquatic plants, seeds, insects, crustaceans, snails, shrimp, bivalves.*
HABITAT *Coast or tundra wetlands while breeding; winters offshore and Great Lakes.*
LENGTH *15–22in (38–56cm)*
WINGSPAN *28–31in (72–79cm)*

Lesser Scaup

S

Aythya affinis

The Lesser Scaup is the most abundant diving duck in North America. The two scaup species are very similar in appearance and are best identified by head shape when stationary. Lesser Scaups generally have a more pointed head than Greater Scaups, but head shape can change with position. For example, the crown feathers are flattened just before diving in both species.

dark wavy pattern on upperparts

narrow head with bump at the rear

purple-green gloss on head

narrow, thin, blue-gray bill

black rear end

♂

pale flanks

black breast and neck

VOICE *Males: mostly silent except wheezy* wheeow wheeow wheeow *during courtship; females: repetitive, grating* garrrf garrrf garrrf.
NESTING *Nest built in tall vegetation, far from water, or on islands and mats of floating vegetation; 8–11 eggs; 1 brood; May–Sep.*
FEEDING *Leeches, crustaceans, mollusks, aquatic insects, aquatic plants and seeds.*
HABITAT *Open northern forests and forest tundra while breeding; coasts, lakes in winter.*
LENGTH *15½–17½in (39–45cm)*
WINGSPAN *27–31in (68–78cm)*

♂

whitish underwings

black head

whitish belly

brown back

white patch around base of gray bill

rich brown head and neck

♀

brown flank feathers with gray fringes

Common Eider

Ⓢ

Somateria mollissima

The largest duck in North America, the Common Eider is also the most numerous, widespread, and variable of the eiders. Four of its seven subspecies occur in North America, and vary in the markings and color of their heads and bills. Male Common Eiders also have considerable seasonal plumage changes, and do not acquire their adult plumage until the third year.

♂❄

whitish underwing

♀

black rump and tail

long, sloping forehead

brown overall

♀

black cap

greenish olive bill

♂❄

white breast, with rose tinge

mottled, black-and-brown upperparts

olive-green wash on nape

VOICE *Repeated hoarse, grating notes korr-korr-korr; males: owl-like ah-WOO-ooo; females: low, gutteral notes krrrr-krrrr-krrrr.*
NESTING *Depression on ground lined with down and plant matter, near water; 2–7 eggs; 1 brood; Jun–Sep.*
FEEDING *Dives in synchronized flocks for mollusks and crustaceans; consumes larger prey above the surface.*
HABITAT *Mostly coastal islands and peninsulas while breeding; coastal waters in winter.*
LENGTH *19½–28in (50–71cm)*
WINGSPAN *31–42in (80–108cm)*

King Eider

S

Somateria spectabilis

The scientific name of the King Eider, *spectabilis*, means "worth seeing," and its gaudy marking and coloring around the head and bill make it hard to mistake. Females resemble the somewhat larger and paler Common Eider. The female King Eider has a more rounded head, more compact body, and a longer bill than the male. King Eiders may dive down to 180ft (55m) when foraging.

white underwing

♂ BREEDING

short neck

long feathers form triangular "sails"

white flank patch

orange to reddish frontal shield, outlined in black

reddish orange bill

pale blue crown and nape

green cheek

rose blush on breast

♂ BREEDING

black underparts

long-billed profile

brown-black upperparts

♀

brown-black upperparts

scalloped breast

"v"-shaped markings on sides

VOICE *Males: repeated series of low, dove-like arrrroooooo, followed by softer cooos when courting; females: grunts and croaks.*
NESTING *Slight depression in tundra lined with vegetation and down; 4–7 eggs; 1 brood; Jun–Sep.*
FEEDING *Dives for mollusks, crustaceans, and starfish; insects and plants when breeding.*
HABITAT *Dry areas near low marshes, islands while breeding; coastal waters in winter.*
LENGTH *18½–25in (47–64cm)*
WINGSPAN *37in (94cm)*

Harlequin Duck

D

Histrionicus histrionicus

This small, hardy duck is a skillful swimmer, diving to forage on the bottom of fast-moving, turbulent streams for its favorite insect prey. Despite the male's unmistakable plumage at close range, it looks very dark from a distance. It can be found among crashing waves, alongside larger and bigger-billed Surf and White-winged Scoters, who feed in the same habitat.

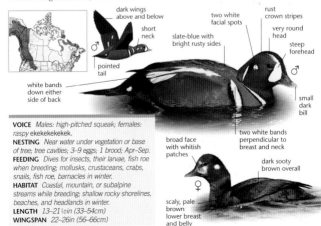

dark wings above and below

short neck

♂

pointed tail

white bands down either side of back

two white facial spots

rust crown stripes

very round head

steep forehead

♂

slate-blue with bright rusty sides

small dark bill

two white bands perpendicular to breast and neck

broad face with whitish patches

dark sooty brown overall

♀

scaly, pale brown lower breast and belly

VOICE *Males: high-pitched squeak; females: raspy ekekekekekek.*
NESTING *Near water under vegetation or base of tree; tree cavities; 3–9 eggs; 1 brood; Apr–Sep.*
FEEDING *Dives for insects, their larvae, fish roe when breeding; mollusks, crustaceans, crabs, snails, fish roe, barnacles in winter.*
HABITAT *Coastal, mountain, or subalpine streams while breeding; shallow rocky shorelines, beaches, and headlands in winter.*
LENGTH *13–21½in (33–54cm)*
WINGSPAN *22–26in (56–66cm)*

Surf Scoter

Melanitta perspicillata ⓢ

Surf Scoters migrate up and down both coasts, often with the other species. They take their name from the way they dive for mollusks through heavy surf. Groups often dive and resurface in unison. Black and Surf Scoters can be difficult to tell apart; look for silvery gray flight feathers on the Black Scoter and black wings overall on the Surf Scoter.

black wings overall

compact body

♂

velvety black feathers

long tail feathers

white eye

white nape

white forehead

large, black spot on bill

♂

swollen, orange bill with white base

all-dark bill

whitish facial patches

dark brown overall

♀

VOICE *Normally silent; courting males: liquid gurgled puk-puk, bubbled whistles, and low croaks; females: harsh, crow-like crahh.*
NESTING *Ground nest lined with down and vegetation on brushy tundra, often under low branches of a conifer tree; 5–10 eggs; 1 brood; May–Sep.*
FEEDING *Mollusks, other aquatic invertebrates.*
HABITAT *Lake islands in forests while breeding; coastal regions in winter.*
LENGTH *19–23½in (48–60cm)*
WINGSPAN *30in (77cm)*

White-winged Scoter

Melanitta fusca ⓢ

The largest of the three scoters, the White-winged Scoter can be identified by its white wing patch. Females look similar to immature Surf Scoters and can be identified by head shape, extent of bill feathering, and shape of white areas on the face. When diving, this scoter leaps forward and up, arching its neck, and opens its wings when entering the water.

white wing patch

appears all-black in flight

upturned white "comma" around white eye

black knob at base of bill

all black with brownish sides

♂

pinkish red to yellow-orange bill

feathers extend onto the bill

dark brown overall

♀

VOICE *Mostly silent; courting males: whistling note; females: growly karr.*
NESTING *Depression lined with twigs and down in dense thickets, often far from water; 8–9 eggs; 1 brood; Jun–Sep.*
FEEDING *Dives for mollusks and crustaceans; sometimes fish and aquatic plants.*
HABITAT *Large freshwater or brackish lakes or ponds while breeding; coasts, bays, and inlets in winter.*
LENGTH *19–23in (48–58cm)*
WINGSPAN *31in (80cm)*

Black Scoter

Melanitta nigra

Black Scoters, the most vocal of the scoters, winter along both coasts of North America. They form dense flocks on the waves, often segregated by gender. While swimming, the Black Scoter sometimes flaps its wings and while doing so, it drops its neck low down, unlike the other two scoters. This scoter breeds in two widely separated sub-Arctic breeding areas.

dark brown eye

black lining on underwings

pale, silvery gray flight feathers

♂

conspicuous yellow-orange knob on black bill

entirely black, heavily built body

black bill with small yellow patch

dark cap

pale brownish gray cheeks

dark brown overall

♀

smaller bill

VOICE Males: high-whistled peeew; females: low raspy kraaa.
NESTING Depression lined with grass and down, often in tall grass on tundra; 5–10 eggs; 1 brood; May–Sep.
FEEDING Dives for mollusks, crustaceans, and plants; aquatic insects, freshwater mussels.
HABITAT Near shallow, small lakes while breeding; shallow water over gravel or sand and offshore ledges in winter.
LENGTH 17–21in (43–53cm)
WINGSPAN 31–35in (79–90cm)

Long-tailed Duck

Clangula hyemalis

The Long-tailed Duck was previously called the Oldsquaw. The male has two extremely long tail feathers, which are often held up in the air. The male's loud calls are quite musical, and, when heard from a flock, have a chorus-like quality. The Long-tailed Duck can dive for a prolonged period of time, and may reach depths of 200ft (60m), making it one of the deepest diving ducks.

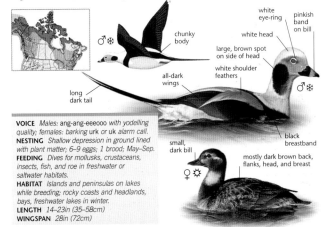

♂❄

chunky body

white eye-ring

pinkish band on bill

white head

large, brown spot on side of head

white shoulder feathers

♂❄

all-dark wings

long dark tail

black breastband

small, dark bill

♀☀

mostly dark brown back, flanks, head, and breast

VOICE Males: ang-ang-eeeooo with yodelling quality; females: barking urk or uk alarm call.
NESTING Shallow depression in ground lined with plant matter; 6–9 eggs; 1 brood; May–Sep.
FEEDING Dives for mollusks, crustaceans, insects, fish, and roe in freshwater or saltwater habitats.
HABITAT Islands and peninsulas on lakes while breeding; rocky coasts and headlands, bays, freshwater lakes in winter.
LENGTH 14–23in (35–58cm)
WINGSPAN 28in (72cm)

Common Goldeneye

(S)

Bucephala clangula

The Common Goldeneye is a medium-sized, compact, diving duck. It closely resembles the Barrow's Goldeneye, to which it is related along with the Bufflehead. It is aggressive and very competitive with members of its own species, as well as other cavity-nesting ducks, regularly laying eggs in the nests of other species. Before diving, the Common Goldeneye flattens its feathers in preparation for underwater foraging. Its wings make a whirring sound in flight.

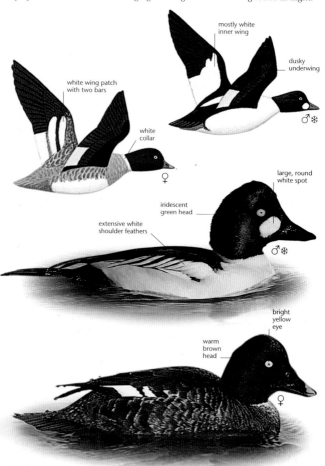

mostly white inner wing

dusky underwing

♂ ❄

white wing patch with two bars

white collar

♀

large, round white spot

iridescent green head

extensive white shoulder feathers

♂ ❄

bright yellow eye

warm brown head

♀

VOICE *Courting males: faint* peent; *females: harsh* gack *or repeated* cuk.
NESTING *Cavity-nester in holes made by other birds, in branches or hollow trees; nest boxes; 4–13 eggs; 1 brood; Apr–Sep.*
FEEDING *Dives during breeding season for insects; in winter, mollusks and crustaceans; fish and plant matter.*
HABITAT *Wetlands, lakes, and rivers in forests while breeding; coastal bays in winter*
LENGTH *15½–20in (40–51cm)*
WINGSPAN *30–33in (77–83cm)*

Bufflehead (S)

Bucephala albeola

The Bufflehead is the smallest diving duck in North America. In flight, males resemble the larger Common Goldeneye, yet the large white area on their head makes them easy to distinguish. The northern limit of the Bufflehead's breeding range corresponds to that of the Northern Flicker, as the ducks usually nest in abandoned Flicker cavities.

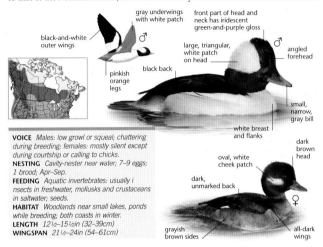

gray underwings with white patch

front part of head and neck has iridescent green-and-purple gloss

black-and-white outer wings

large, triangular, white patch on head

angled forehead

pinkish orange legs

black back

small, narrow, gray bill

white breast and flanks

dark brown head

oval, white cheek patch

dark, unmarked back

grayish brown sides

all-dark wings

VOICE *Males: low growl or squeal; chattering during breeding; females: mostly silent except during courtship or calling to chicks.*
NESTING *Cavity-nester near water; 7–9 eggs; 1 brood; Apr–Sep.*
FEEDING *Aquatic invertebrates: usually insects in freshwater, mollusks and crustaceans in saltwater; seeds.*
HABITAT *Woodlands near small lakes, ponds while breeding; both coasts in winter.*
LENGTH 12½–15½in (32–39cm)
WINGSPAN 21½–24in (54–61cm)

Barrow's Goldeneye (D)

Bucephala islandica

Barrow's Goldeneye is a slightly larger, darker version of the Common Goldeneye. The bill color varies seasonally and geographically: Eastern Barrow's have blacker bills with less yellow, and western populations have entirely yellow bills, which darken in summer. During the breeding season, the majority of Barrow's Goldeneyes are found in mountainous regions of northwest North America.

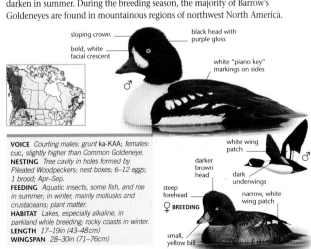

sloping crown

black head with purple gloss

bold, white facial crescent

white "piano key" markings on sides

white wing patch

darker brown head

dark underwings

steep forehead

narrow, white wing patch

♀ BREEDING

small, yellow bill

VOICE *Courting males: grunt ka-KAA; females: cuc, slightly higher than Common Goldeneye.*
NESTING *Tree cavity in holes formed by Pileated Woodpeckers; nest boxes; 6–12 eggs; 1 brood; Apr–Sep.*
FEEDING *Aquatic insects, some fish, and roe in summer; in winter, mainly mollusks and crustaceans; plant matter.*
HABITAT *Lakes, especially alkaline, in parkland while breeding; rocky coasts in winter.*
LENGTH 17–19in (43–48cm)
WINGSPAN 28–30in (71–76cm)

Hooded Merganser (S)

Lophodytes cucullatus

The smallest of the three mergansers, Hooded Mergansers have crests that they can raise or flatten. The male's raised crest displays a gorgeous white fan, surrounded by black. The Hooded Merganser and the Wood Duck can be confused when seen in flight since they both are fairly small with bushy heads and long tails.

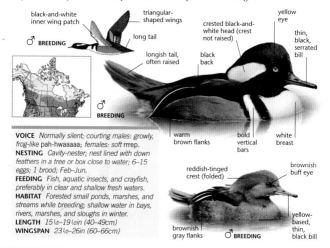

black-and-white inner wing patch
triangular-shaped wings
♂ BREEDING
long tail
crested black-and-white head (crest not raised)
yellow eye
thin, black, serrated bill
longish tail, often raised
black back
♂ BREEDING
black back
warm brown flanks
bold vertical bars
white breast

reddish-tinged crest (folded)
brownish buff eye
brownish gray flanks
♂ BREEDING
yellow-based, thin, black bill

VOICE *Normally silent; courting males: growly, frog-like pah-hwaaaaa; females: soft rrrep.*
NESTING *Cavity-nester; nest lined with down feathers in a tree or box close to water; 6–15 eggs; 1 brood; Feb–Jun.*
FEEDING *Fish, aquatic insects, and crayfish, preferably in clear and shallow fresh waters.*
HABITAT *Forested small ponds, marshes, and streams while breeding; shallow water in bays, rivers, marshes, and sloughs in winter.*
LENGTH *15½–19¼in (40–49cm)*
WINGSPAN *23½–26in (60–66cm)*

Common Merganser (S)

Mergus merganser

The Common Merganser is the largest of the three merganser species in North America. This large fish-eater is common and widespread, particularly in the northern portion of its range. It is often found in big flocks on lakes or smaller groups along rivers. It spends most of its time on the water, using its serrated bill to catch fish underwater.

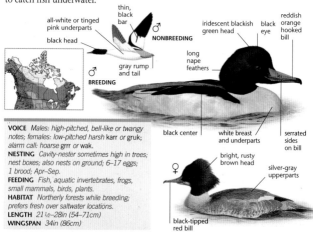

all-white or tinged pink underparts
thin, black bar
♂ NONBREEDING
black head
iridescent blackish green head
black eye
reddish orange hooked bill
long nape feathers
gray rump and tail
♂ BREEDING
black center
white breast and underparts
serrated sides on bill
bright, rusty brown head
♀
silver-gray upperparts
black-tipped red bill

VOICE *Males: high-pitched, bell-like or twangy notes; females: low-pitched harsh karr or gruk; alarm call: hoarse grrr or wak.*
NESTING *Cavity-nester sometimes high in trees; nest boxes; also nests on ground; 6–17 eggs; 1 brood; Apr–Sep.*
FEEDING *Fish, aquatic invertebrates, frogs, small mammals, birds, plants.*
HABITAT *Northerly forests while breeding; prefers fresh over saltwater locations.*
LENGTH *21½–28in (54–71cm)*
WINGSPAN *34in (86cm)*

Red-breasted Merganser

S

Mergus serrator

The Red-breasted Merganser is somewhat smaller than the Common Merganser but larger than the Hooded. Both sexes are easily recognized by their long, sparse, ragged-looking double crest. The Red-breasted Merganser, unlike the other two mergansers, nests on the ground, in loose colonies, often among gulls and terns, and is protected by its neighbors.

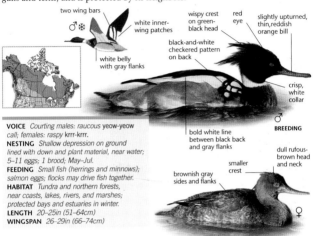

♂ ❄
two wing bars
white inner-wing patches
white belly with gray flanks

wispy crest on green-black head
red eye
slightly upturned, thin, reddish orange bill
black-and-white checkered pattern on back
crisp, white collar

♂
BREEDING
bold white line between black back and gray flanks

dull rufous-brown head and neck
smaller crest
brownish gray sides and flanks
♀

VOICE Courting males: raucous yeow-yeow call; females: raspy krrr-krrr.
NESTING Shallow depression on ground lined with down and plant material, near water; 5–11 eggs; 1 brood; May–Jul.
FEEDING Small fish (herrings and minnows); salmon eggs; flocks may drive fish together.
HABITAT Tundra and northern forests, near coasts, lakes, rivers, and marshes; protected bays and estuaries in winter.
LENGTH 20–25in (51–64cm)
WINGSPAN 26–29in (66–74cm)

Ruddy Duck

S

Oxyura jamaicensis

Ruddy Ducks often hold their tails cocked upward, especially when sleeping. During courtship displays, the male points its long tail skyward while rapidly thumping its electric blue bill against its chest. The Ruddy Duck is an excellent swimmer and diver but very awkward on land; females are known to push themselves along instead of walking.

broad, short wings with whitish wing linings
♂ BREEDING
pale belly

black cap and nape
large head
bright blue bill, slightly knobby at base

long tail, often erect
large, white cheek patches
♂
BREEDING

rich cinnamon body and neck
arched dark line on cheek
dark bill
brownish upperparts
♀
paler flanks

VOICE Females: nasal raanh, high pitched eeek; males: popping noises with feet.
NESTING Platform, bowl-shaped nest built over water in thick vegetation, rarely on land; 6–10 eggs; 1 brood; May–Sep.
FEEDING Aquatic insects, larvae, crustaceans, and other invertebrates, particularly when breeding; also plants during winter.
HABITAT Shallow wetlands while breeding; mostly freshwater habitats in winter.
LENGTH 14–17in (35–43cm)
WINGSPAN 22–24in (56–62cm)

Loons

Loons are almost entirely aquatic birds; their legs are positioned so far to the rear of their body that they must shuffle on their bellies when they go from water to land. In summer they are found on rivers, lakes, and ponds, where they nest close to the water's edge. After breeding, they occur along coasts, often after flying hundreds of miles away from their freshwater breeding grounds. Excellent swimmers and divers, loons are unusual among birds in that their bones are less hollow than those of other groups. Consequently, they can expel air from their lungs and compress their body feathers until they slowly sink beneath the surface. They can remain submerged for several minutes.

FEEDING TIME
A Red-throated Loon gives a fish to its chick to gulp down headfirst and whole.

Red-throated Loon

Gavia stellata

This elegant loon is almost unmistakable, with a pale, slim body, upward tilted head, and a thin, upturned bill. Unlike other loons, the Red-throated Loon can launch straight into the air from both land and water, but most of the time it tends to use a "runway." In an elaborate breeding ritual, a pair of birds races side by side upright across the surface of water.

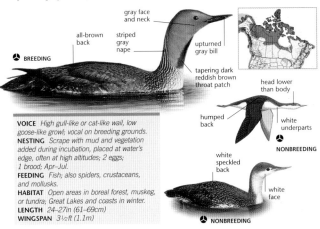

BREEDING

all-brown back

striped gray nape

gray face and neck

upturned gray bill

tapering dark reddish brown throat patch

head lower than body

humped back

white underparts

NONBREEDING

white speckled back

white face

NONBREEDING

VOICE *High gull-like or cat-like wail, low goose-like growl; vocal on breeding grounds.*
NESTING *Scrape with mud and vegetation added during incubation, placed at water's edge, often at high altitudes; 2 eggs; 1 brood; Apr–Jul.*
FEEDING *Fish; also spiders, crustaceans, and mollusks.*
HABITAT *Open areas in boreal forest, muskeg, or tundra; Great Lakes and coasts in winter.*
LENGTH *24–27in (61–69cm)*
WINGSPAN *3½ft (1.1m)*

Pacific Loon

Gavia pacifica

Although the Pacific Loon's breeding range is smaller than the Common Loon's, it is believed to be the most abundant loon species in North America. This conspicuous migrant along the Pacific Coast in spring shares its habitat in northern Alaska with the slightly larger and darker Arctic Loon. The Pacific Loon can remain underwater for especially sustained periods of time, usually in pursuit of fish.

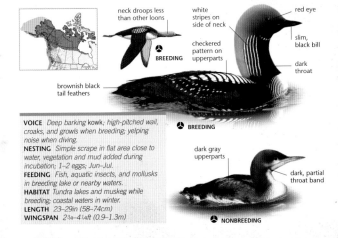

neck droops less than other loons

BREEDING

white stripes on side of neck

checkered pattern on upperparts

red eye

slim, black bill

dark throat

brownish black tail feathers

BREEDING

dark gray upperparts

dark, partial throat band

NONBREEDING

VOICE *Deep barking kowk; high-pitched wail, croaks, and growls when breeding; yelping noise when diving.*
NESTING *Simple scrape in flat area close to water, vegetation and mud added during incubation; 1–2 eggs; Jun–Jul.*
FEEDING *Fish, aquatic insects, and mollusks in breeding lake or nearby waters.*
HABITAT *Tundra lakes and muskeg while breeding; coastal waters in winter.*
LENGTH *23–29in (58–74cm)*
WINGSPAN *2¾–4¼ft (0.9–1.3m)*

Common Loon

S

Gavia immer

The Common Loon is the provincial bird of Ontario and has the largest range of all loons in North America. It is slightly smaller than the Yellow-billed Loon but larger than the other three loons. It can remain underwater for well over 10 minutes, although it usually stays submerged for 40 seconds to 2 minutes while fishing. Occasionally, it interbreeds with the Yellow-billed, Arctic, or Pacific Loon.

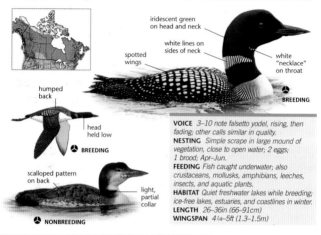

iridescent green on head and neck

white lines on sides of neck

spotted wings

white "necklace" on throat

BREEDING

humped back

head held low

BREEDING

scalloped pattern on back

light, partial collar

NONBREEDING

VOICE *3–10 note falsetto yodel, rising, then fading; other calls similar in quality.*
NESTING *Simple scrape in large mound of vegetation, close to open water; 2 eggs; 1 brood; Apr–Jun.*
FEEDING *Fish caught underwater; also crustaceans, mollusks, amphibians, leeches, insects, and aquatic plants.*
HABITAT *Quiet freshwater lakes while breeding; ice-free lakes, estuaries, and coastlines in winter.*
LENGTH *26–36in (66–91cm)*
WINGSPAN *4¼–5ft (1.3–1.5m)*

Yellow-billed Loon

S

Gavia adamsii

The largest of the loons, the Yellow-billed Loon has the most restricted range and smallest global population. About three quarters of the estimated 16,000 birds live in North America. It makes the most of the short nesting season, arriving at its breeding grounds already paired and breeding immediately, although extensive ice formation can prevent it from breeding in some years.

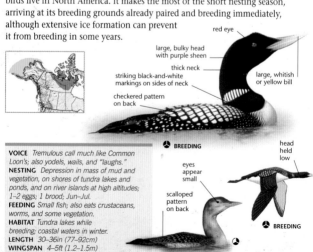

red eye

large, bulky head with purple sheen

thick neck

striking black-and-white markings on sides of neck

large, whitish or yellow bill

checkered pattern on back

BREEDING

head held low

eyes appear small

scalloped pattern on back

BREEDING

VOICE *Tremulous call much like Common Loon's; also yodels, wails, and "laughs."*
NESTING *Depression in mass of mud and vegetation, on shores of tundra lakes and ponds, and on river islands at high altitudes; 1–2 eggs; 1 brood; Jun–Jul.*
FEEDING *Small fish; also eats crustaceans, worms, and some vegetation.*
HABITAT *Tundra lakes while breeding; coastal waters in winter.*
LENGTH *30–36in (77–92cm)*
WINGSPAN *4–5ft (1.2–1.5m)*

Tubenoses

The tubenoses are characterized by the tubular nostrils for which they are named. These nostrils act like water pistols to help to get rid of excess salt, and may enhance sense of smell.

The long, narrow wings of albatrosses are perfectly suited for almost endless flight in the strong, constant ocean air currents. Although expert gliders, they need lift from wind to take off from the ground.

Smaller than albatrosses, shearwaters and gadfly petrels range over all the world's oceans, especially the Pacific. During and after storms are the best times to look for these birds, as this is when they have been drifting away from the deep sea due to wind and waves.

The smallest tubenoses in North American waters, the storm-petrels are also the most agile fliers. They spend most of their lives flying over the open sea, only visiting land in the breeding season, when they form huge colonies.

STRONG PAIR BOND
After elaborate courtship displays, albatrosses generally pair for life.

Black-footed Albatross (D)

Phoebastria nigripes

The most frequently seen albatross in North American waters, this distinctive all-dark bird breeds mainly on the Hawaiian Islands and regularly visits the Pacific Coast during the nonbreeding season. Unfortunately, a tendency to scavenge around fishing boats results in this and other species of albatross being drowned when they are accidentally hooked on long lines or tangled in drift nets.

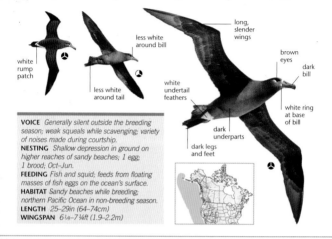

white rump patch

less white around bill

less white around tail

long, slender wings

brown eyes

dark bill

white undertail feathers

dark underparts

white ring at base of bill

dark legs and feet

VOICE *Generally silent outside the breeding season; weak squeals while scavenging; variety of noises made during courtship.*
NESTING *Shallow depression in ground on higher reaches of sandy beaches; 1 egg; 1 brood; Oct–Jun.*
FEEDING *Fish and squid; feeds from floating masses of fish eggs on the ocean's surface.*
HABITAT *Sandy beaches while breeding; northern Pacific Ocean in non-breeding season.*
LENGTH *25–29in (64–74cm)*
WINGSPAN *6¼–7¼ft (1.9–2.2m)*

Northern Fulmar (S)

Fulmarus glacialis

Possessing paddle-shaped wings and color patterns ranging from almost all-white to all-gray, the Northern Fulmar breeds at high latitudes, then disperses south to offshore waters on both coasts. The Northern Fulmar can often be seen in large mixed flocks containing albatrosses, shearwaters, and petrels. Fulmars often follow fishing boats, eager to pounce on the offal thrown overboard.

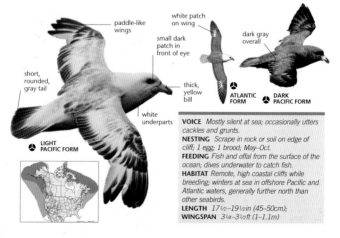

paddle-like wings

white patch on wing

small dark patch in front of eye

dark gray overall

short, rounded, gray tail

thick, yellow bill

white underparts

ATLANTIC FORM

DARK PACIFIC FORM

LIGHT PACIFIC FORM

VOICE *Mostly silent at sea; occasionally utters cackles and grunts.*
NESTING *Scrape in rock or soil on edge of cliff; 1 egg; 1 brood; May–Oct.*
FEEDING *Fish and offal from the surface of the ocean; dives underwater to catch fish.*
HABITAT *Remote, high coastal cliffs while breeding; winters at sea in offshore Pacific and Atlantic waters, generally further north than other seabirds.*
LENGTH *17½–19½in (45–50cm);*
WINGSPAN *3¼–3½ft (1–1.1m)*

Cory's Shearwater

Ⓢ

Calonectris diomedea

Studies of Cory's Shearwaters off the Atlantic coast suggest the presence of two forms. The more common form, *C. d. borealis*, nests in the eastern Atlantic and is chunkier, with less white in the wing from below. The other form, *C. d.diomedea*, breeds in the Mediterranean, has a more slender build and more extensive white under the wing.

scalloped pattern

grayish head and chin

yellow bill with dark tip

dark wingtip and trailing edge

white breast, with sooty-gray sides

all white belly

clean white underwing

long, pointed wings

pale rump

VOICE *Mostly silent at sea; descending, lamb-like bleating.*
NESTING *Nests in burrow or rocky crevice; 1 egg; 1 brood; May–Sep.*
FEEDING *Dives into water or picks at surface for small schooling fish, and marine invertebrates such as squid.*
HABITAT *Breeds mostly in the Mediterranean and on islands of the eastern Atlantic; winters widely over the Atlantic Ocean.*
LENGTH *18in (46cm)*
WINGSPAN *3½ft (1.1m)*

Manx Shearwater

Ⓢ

Puffinus puffinus

Most shearwaters are little known, but the Manx is an exception. It is common in the British Isles, and ornithologists have been studying it there for decades. Long-term banding programs revealed one bird that flew over 3,000 miles (4,800km) from Massachusetts to its nesting burrow in Wales in just 12½ days, and another that was captured 56 years after it was first banded.

dark upperwings

short tail

small head

black edge of wing

dark, hooked bill

white throat

white undertail feathers

snow white underparts

long, pointed wings

crisp white underwings

VOICE *Usually silent at sea; loud and raucous kah-kah-kah-kah-kah-HOWW at breeding sites.*
NESTING *In burrow, in peaty soil, or rocky crevice; 1 egg; 1 brood; Apr–Oct.*
FEEDING *Dives into water, often with open wings and stays underwater, or picks at surface for small schooling fish and squid.*
HABITAT *Islands in eastern North Atlantic while breeding; rarely seen from shore during migration and winter.*
LENGTH *13½in (34cm)*
WINGSPAN *33in (83cm)*

Sooty Shearwater

(S)

Puffinus griseus

Sooty Shearwaters are extremely long-distance migrants. Pacific populations in particular travel as far as 300 miles (480km) per day and an extraordinary 45,000 miles (72,500km) or more per year. It is fairly easy to identify off the East Coast of North America, as it is the only all-dark shearwater found there.

all-dark underparts

silvery white patch along underwing

long, slender wings

sooty head

all-dark upperparts

long, hooked bill

VOICE *Silent at sea; occasionally gives varied, agitated vocalizations when feeding, very loud calls at breeding colonies.*
NESTING *In burrow or rocky crevice; 1 egg; 1 brood; Oct–May.*
FEEDING *Dives and picks at surface for small schooling fish and mollusks such as squid.*
HABITAT *Islands in southern oceans while breeding; open ocean off both coasts in non-breeding season.*
LENGTH *18in (46cm)*
WINGSPAN *3ft 3in (1m)*

Pink-footed Shearwater

(T)

Puffinus creatopus

In many ways, this species is the West Coast equivalent of Cory's Shearwater. The way it holds its wings (angled at the "wrist"), its size, and its flight style are all reminiscent of Cory's. Though Pink-footed Shearwaters are fairly variable in plumage, they are always rather dull, with little color variation. However, it can be distinguished by its distinctly pinkish bill.

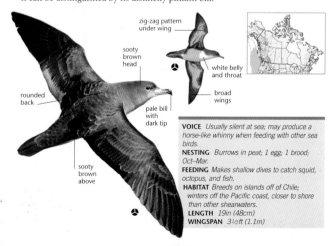

zig-zag pattern under wing

sooty brown head

white belly and throat

broad wings

rounded back

pale bill with dark tip

sooty brown above

VOICE *Usually silent at sea; may produce a horse-like whinny when feeding with other sea birds.*
NESTING *Burrows in peat; 1 egg; 1 brood; Oct–Mar.*
FEEDING *Makes shallow dives to catch squid, octopus, and fish.*
HABITAT *Breeds on islands off of Chile; winters off the Pacific coast, closer to shore than other shearwaters.*
LENGTH *19in (48cm)*
WINGSPAN *3½ft (1.1m)*

Greater Shearwater (S)

Puffinus gravis

The Greater Shearwater is similar in size to Cory's Shearwater and the birds scavenge together for scraps around fishing boats. However, while the Cory's Shearwater has slow, labored wing beats, and glides high on broad, swept-back wings, Greater Shearwaters keep low, flapping hurriedly between glides on straight, narrow wings. The brown smudges on the belly and paler underwings of the Greater Shearwater also help distinguish the species.

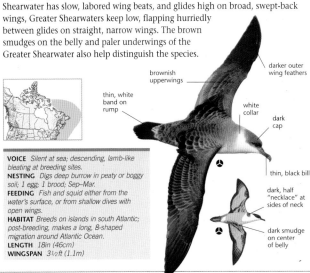

brownish upperwings

darker outer wing feathers

thin, white band on rump

white collar

dark cap

thin, black bill

dark, half "necklace" at sides of neck

dark smudge on center of belly

VOICE *Silent at sea; descending, lamb-like bleating at breeding sites.*
NESTING *Digs deep burrow in peaty or boggy soil; 1 egg; 1 brood; Sep–Mar.*
FEEDING *Fish and squid either from the water's surface, or from shallow dives with open wings.*
HABITAT *Breeds on islands in south Atlantic; post-breeding, makes a long, 8-shaped migration around Atlantic Ocean.*
LENGTH *18in (46cm)*
WINGSPAN *3½ft (1.1m)*

Wilson's Storm-Petrel (S)

Oceanites oceanicus

Wilson's Storm-Petrel is the quintessential small oceanic petrel. After breeding in the many millions on the Antarctic Peninsula and various islands there, it moves north to spend the summer off North America's Atlantic coast and be a familiar sight to fishermen and birders at sea. By August they can be seen lingering, but by October they have flown south.

dark wings and body

small, black "tube nose"

yellow webbing between toes

pale bar on upperwing

short, square tail

white rump and lower flanks

broad, pointed wings

"walking" on water

VOICE *At sea, soft rasping notes; variety of coos, churrs, and twitters during the night at breeding sites.*
NESTING *In rock crevices; also burrows where there is peaty soil; 1 egg; 1 brood; Nov–Mar.*
FEEDING *Patters on the water's surface, legs extended, picking up tiny crustaceans; also carrion, droplets of oil.*
HABITAT *Rocky islets, cliffs, and boulder scree while breeding; Apr–Sep off the Atlantic coasts.*
LENGTH *6¾in (17cm)*
WINGSPAN *16in (41cm)*

Leach's Storm-Petrel   ⓢ

Oceanodroma leucorhoa

Leach's Storm-Petrel is widespread in both the Atlantic and Pacific Oceans. It breeds in colonies on islands off the coasts, coming to land at night and feeding offshore during the day, often many miles from the colony. This storm-petrel has both geographical and individual variation; most populations show a white rump, but others have a rump that is the same color as the rest of the body.

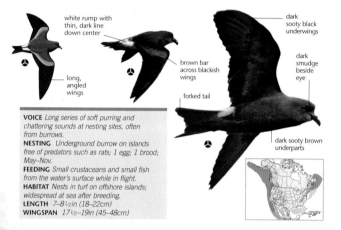

white rump with thin, dark line down center

long, angled wings

dark sooty black underwings

brown bar across blackish wings

forked tail

dark smudge beside eye

dark sooty brown underparts

VOICE *Long series of soft purring and chattering sounds at nesting sites, often from burrows.*
NESTING *Underground burrow on islands free of predators such as rats; 1 egg; 1 brood; May–Nov.*
FEEDING *Small crustaceans and small fish from the water's surface while in flight.*
HABITAT *Nests in turf on offshore islands; widespread at sea after breeding.*
LENGTH *7–8½in (18–22cm)*
WINGSPAN *17½–19in (45–48cm)*

Fork-tailed Storm-Petrel   ⓢ

Oceanodroma furcata

The Fork-tailed Storm-Petrel is distinctive with its ghostly silvery gray plumage and forked tail. It is the most northerly breeding storm-petrel in the North Pacific, nesting all the way north to the Aleutian Islands. It incubates its eggs at lower temperatures than other petrels do, and its nestlings can be left alone between feedings for a longer time because they can conserve energy by lowering their body temperature.

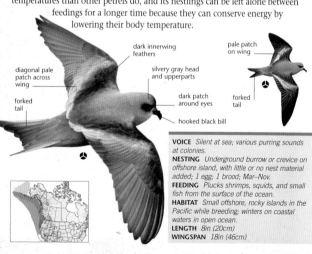

dark innerwing feathers

pale patch on wing

diagonal pale patch across wing

silvery gray head and upperparts

forked tail

dark patch around eyes

hooked black bill

forked tail

VOICE *Silent at sea; various purring sounds at colonies.*
NESTING *Underground burrow or crevice on offshore island, with little or no nest material added; 1 egg; 1 brood; Mar–Nov.*
FEEDING *Plucks shrimps, squids, and small fish from the surface of the ocean.*
HABITAT *Small offshore, rocky islands in the Pacific while breeding; winters on coastal waters in open ocean.*
LENGTH *8in (20cm)*
WINGSPAN *18in (46cm)*

Grebes

Grebes resemble loons and share many of their aquatic characteristics, including a streamlined shape and legs that are positioned far back on the body, making them awkward on land. Grebe nests are usually partially floating platforms, built on beds of water plants.

They dive to catch fish with a short, forward arching spring, using their lobed toes in a sideways motion to propel them underwater. Unusually among birds, they swallow feathers, supposedly to trap fish bones and protect their stomachs, then periodically disgorge them. Like loons, grebes can control their buoyancy by exhaling air and compressing their plumage so that they sink quietly below the surface. They are strong fliers, and migratory.

A FINE DISPLAY
A Horned Grebe reveals colorful plumes in its courtship display.

Pied-billed Grebe

S

Podilymbus podiceps

The widest ranging of the North American grebes, the Pied-billed Grebe is tolerant of populated areas and breeds on lakes and ponds. A powerful swimmer, it can remain submerged for 16–30 seconds when it dives. Its courtship ritual is more vocal than visual; pairs usually duet-call in the mating season. Migration begins when its breeding area ices up and food becomes scarce. The Pied-billed Grebe is capable of sustained flights of over 2,000 miles (3,200km).

outstretched neck

BREEDING

lighter flight feathers

yellowish bill

whitish throat

NONBREEDING

whitish, hooked bill with a black ring

brown eye

reddish brown neck and breast

brownish gray body

black throat patch

BREEDING

white undertail

VOICE *Grunts and wails; in spring, cuckoo-like repeated gobble kup-kup-Kaow-Kaow-kaow.*
NESTING *Floating nest of partially decayed plants and clipped leaves, attached to vegetation in marshes and quiet waters; 4–7 eggs; 2 broods; Apr–Oct.*
FEEDING *Dives for crustaceans, fish, amphibians, and insects; picks prey from emergent vegetation, or catches them mid-air.*
HABITAT *Coastal brackish ponds, seasonal ponds, and marshes.*
LENGTH *12–15in (31–38cm);*
WINGSPAN *18–24in (46–62cm)*

Red-necked Grebe ⓢ

Podiceps grisegena

The Red-necked Grebe migrates over short to medium distances and spends the winter along both coasts, where large flocks may be seen during the day. It runs along the water's surface to become airborne, although it rarely flies. This grebe does not come ashore often; it stands erect, but walks awkwardly, and prefers to sink to its breast and shuffle along.

white-edged inner wing

🦅 BREEDING

head and neck in line with body

gray flanks

broad head with crest at rear

grayish white cheeks and throat

black cap

chestnut brown neck and chest

brown eye

🦅 BREEDING

brownish cap

pale, reddish brown crescent near ear

🦅 NONBREEDING

VOICE Nasal, gull-like call on breeding grounds, evolves into bray, ends with whinny; also honks, rattles, hisses, purrs, and ticks.
NESTING Compact, buoyant mound of vegetation in sheltered, shallow marshes and lakes; 4–5 eggs; 1 brood; May–Jul.
FEEDING Fish, crustaceans, aquatic insects, worms, mollusks, salamanders, and tadpoles.
HABITAT Large marshes and small lakes while breeding; estuaries, inlets, bays in winter.
LENGTH 16½–22in (42–56cm)
WINGSPAN 24–35in (61–88cm)

Eared Grebe ⓢ

Podiceps nigricollis

The abundant Eared Grebe is remarkable in terms of physiology. After breeding, it undergoes a complex reorganization of body-fat stores, along with changes in muscle, heart, and digestive organ mass to prepare it for fall migration. All of this increases the bird's energy reserves and body mass, but renders it flightless for up to 10 months.

white patch on wing

out-stretched neck

🦅☼

dark back

🦅☼

black crest

large, wispy gold patch behind red eye

black neck

red eye

thin, upturned bill

rufous breast and sides

dusky cheek

dusky white flanks

🦅❄

upturned bill

grayish neck

VOICE Squeaky, rising poo-eep during courtship; sharp chirp when alarmed; usually silent at other times.
NESTING Sodden nest of decayed plants in reeds in shallow water of marshes, ponds, and lakes; 1 brood; 1–8 eggs; May–Jul.
FEEDING Small crustaceans and aquatic insects; small fish, mollusks; worms in winter.
HABITAT Marshes, shallow lakes, and ponds while breeding; high-saline waters in winter.
LENGTH 12–14in (30–35cm)
WINGSPAN 22½–24in (57–62cm)

Horned Grebe

Podiceps auritus

The timing of the Horned Grebe's migration depends largely on the weather—this species may not leave until its breeding grounds get iced over, nor does it arrive before the ice melts. Its breeding behavior is well documented since it is approachable on nesting grounds and has an elaborate breeding ritual. This grebe's so-called "horns" are in fact yellowish feather patches located behind its eyes, which it can raise at will.

neck and head in line with body

flattish top of head

white cheek

white sides to neck

gold streak from eye to nape

black crown

red eye

short, dark bill with whitish tip

rufous neck

black throat

MOLT

VOICE Descending *aaanrrh* call most common in winter, ends in trill; muted conversational calls when in groups.
NESTING Floating, soggy nest, hidden in vegetation, in small ponds and lake inlets; 3–9 eggs; 1 brood; May–Jul.
FEEDING Dives in open water or forages among plants for small crustaceans, insects, leeches, mollusks, amphibians, fish, and plants.
HABITAT Freshwater or brackish water with emergent vegetation while breeding; coastlines or large bodies of freshwater in winter.
LENGTH 12–15in (30–38cm)
WINGSPAN 18–24in (46–62cm)

Western Grebe

Aechmophorus occidentalis

Western and Clark's Grebes are strictly North American species, sharing much of their breeding habitat and elaborate mating rituals. Interbreeding is uncommon, perhaps because of slight differences in calls, bill colors, and facial patterns. Female Western Grebes are smaller than males and have smaller, thinner, slightly upturned bills. The Western Grebe dives more frequently than Clark's.

whitish band on dark wing

black crown extends below eye

distinctive red eye

dark gray back

black nape stripe

long, slender, slightly upturned greenish yellow bill

brilliant white throat, breast, and belly

dark patch around eyes

light gray back

light white-gray neck

VOICE Alarm, begging, and mating calls; advertising call a harsh, rolling two-noted *krrrikk krrreek.*
NESTING Floating pile of plants, attached to submerged vegetation; occasionally on land; 2–3 eggs; 1 brood; May–Jul.
FEEDING Freshwater or saltwater fish; also crustaceans, worms, occasionally insects.
HABITAT Freshwater lakes and marshes with emergent vegetation; bays, estuaries in winter.
LENGTH 21½–30in (55–75cm)
WINGSPAN 30–39in (76–100cm)

Clark's Grebe

Aechmophorus clarkii

Clark's and Western grebes are closely related and very difficult to distinguish. They rarely fly except when migrating at night. Both species seldom come to land, where their movement is awkward. Their flight muscles suffer wastage after their arrival on the breeding grounds, but during the incubation period adults may feed several miles from the colony by following continuous water trails.

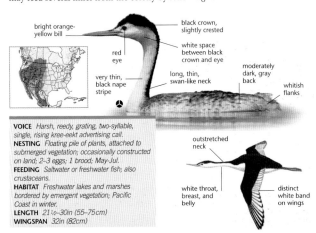

bright orange-yellow bill

black crown, slightly crested

red eye

white space between black crown and eye

very thin, black nape stripe

long, thin, swan-like neck

moderately dark, gray back

whitish flanks

VOICE Harsh, reedy, grating, two-syllable, single, rising *kree-eekt* advertising call.
NESTING Floating pile of plants, attached to submerged vegetation; occasionally constructed on land; 2–3 eggs; 1 brood; May–Jul.
FEEDING Saltwater or freshwater fish; also crustaceans.
HABITAT Freshwater lakes and marshes bordered by emergent vegetation; Pacific Coast in winter.
LENGTH 21½–30in (55–75cm)
WINGSPAN 32in (82cm)

outstretched neck

white throat, breast, and belly

distinct white band on wings

Ibises & Herons

Birds of the waterside or dry land, ibises are characterized by rounded bodies, medium-long legs, short tails, rounded wings, and small, often bare, heads on curved necks, merging into long, curved bills. These gregarious birds fly in long lines or "V" formation and feed mostly on insects, worms, small mollusks, and crustaceans, probing for them in the water and wet mud.

Herons, bitterns, and egrets are mostly waterside birds with long, slender toes, broad, rounded wings, very short tails, forward-facing eyes, and dagger-shaped bills. Bitterns and night-herons have a shawl of smooth, dense neck feathers, while an egret's long, slender neck is tightly feathered, with an obvious "kink" that allows a lightning-fast stab for prey. Bitterns, herons, and egrets fly with their legs trailing and their necks coiled back into their shoulders. Some make obvious bulky treetop nests and feed in the open, while others, especially bitterns, nest and feed secretively.

DANCING ON AIR
The Great Egret's courtship display often involves spreading its wings and leaping in an aerial dance.

Glossy Ibis Ⓢ

Plegadis falcinellus

Glossy Ibises are dark, long-legged, and have a long, curved bill.
Well known for their wandering tendencies, they can also be
found in southern Europe, Asia, Australia, and Africa.
Confined to Florida until the mid-20th century, this
species then started spreading northward, eventually
as far as southern Ontario and across into Nova Scotia.

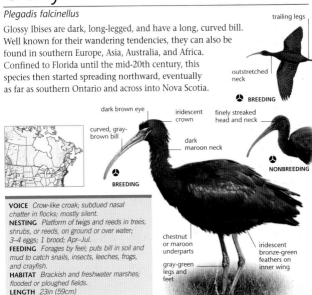

VOICE *Crow-like croak; subdued nasal chatter in flocks; mostly silent.*
NESTING *Platform of twigs and reeds in trees, shrubs, or reeds, on ground or over water; 3–4 eggs; 1 brood; Apr–Jul.*
FEEDING *Forages by feel; puts bill in soil and mud to catch snails, insects, leeches, frogs, and crayfish.*
HABITAT *Brackish and freshwater marshes; flooded or ploughed fields.*
LENGTH *23in (59cm)*
WINGSPAN *36in (92cm)*

American Bittern Ⓢ

Botaurus lentiginosus

The American Bittern's camouflaged plumage and secretive behavior help it
to blend into the thick vegetation of its freshwater wetland habitat. It is heard
much more often than it is seen; its call is unmistakable and has given rise
to many evocative colloquial names, such as "thunder pumper."

VOICE *Deep, resonant pump-er-unk, pump-er-unk; calls mainly at dawn, dusk, and night.*
NESTING *Platform or mound constructed of marsh vegetation, usually over shallow water; 2–7 eggs; 1 brood; Apr–Aug.*
FEEDING *Strikes downward with bill to catch fish, insects, crustaceans, snakes, amphibians, and small mammals.*
HABITAT *Heavily vegetated freshwater wetlands while breeding.*
LENGTH *23½–31in (60–80cm)*
WINGSPAN *3½–4¼ft (1.1–1.3m)*

Least Bittern

Ixobrychus exilis

The smallest heron in North America, the Least Bittern is also one of the most colorful, but its secretive nature makes it easy to overlook in its densely vegetated marsh habitat. A dark color form, which was originally described in the 1800s as a separate species named Cory's Bittern, has rarely been reported in recent decades.

black back

♂

black cap

buff and black pattern on wings

♂

short tail

brown streaks on chest

long, yellow bill

yellowish legs and toes

dark-brown cap

dark-brown back

pale wing feathers

♀

VOICE *Soft ku, ku, ku, ku, ku display call; year-round, a loud kak, kak, kak.*
NESTING *Platform of marsh vegetation and sticks, usually close to open water; 2–7 eggs; 1 brood; Apr–Aug.*
FEEDING *Small fish, insects including dragonflies; also crustaceans; clings quietly to vegetation before striking prey, or stalks slowly.*
HABITAT *Lowland freshwater and rarely brackish marshes while breeding.*
LENGTH *11–14in (28–36cm)*
WINGSPAN *15½–18in (40–46cm)*

Black-crowned Night-Heron

Nycticorax nycticorax

The Black-crowned Night-Heron is chunky and squat. It is also one of the most common and widespread herons in the world, but because it is mainly active at twilight and at night, many people have never seen one. However, its distinctive barking call can be heard at night—even at the center of large cities.

gray wings

long, white head plumes

white spots on brown back

pale lower bill

black back

black crown

short, thick bill

short neck

yellow legs; red in spring

VOICE *Loud, distinctive quark or wok, often given in flight and around colonies.*
NESTING *Large stick nests built usually 20–40ft (6–12m) up in trees; 3–5 eggs; 1 brood; Nov–Aug.*
FEEDING *Aquatic animals (fish, crustaceans, insects, and mollusks); also eggs and chicks of colonial birds (egrets, ibises, and terns).*
HABITAT *Waterbodies, including lakes, ponds, streams, and marshes.*
LENGTH *23–26in (58–65cm)*
WINGSPAN *3½–4ft (1.1–1.2m)*

Green Heron

Butorides virescens

A small, solitary, and secretive bird of dense thicketed wetlands, the Green Heron can be difficult to observe. This dark, crested heron is most often seen flying away from a perceived threat, emitting a loud squawk. While the Green Heron of North and Central America has now been recognized as a separate species, it was earlier grouped with the Green-backed Heron *(B. striatus)*, which is found in the tropics and subtropics throughout the world.

greenish back

BREEDING

long back plumes

BREEDING

glossy orange legs

white speckles on wings

paler bill

greenish black cap

short, rufous neck

white chin

cream streak extends from throat to belly

yellowish legs and feet

NONBREEDING

VOICE Squawking keow when flying from disturbance.
NESTING Nest of twigs in bushes or trees, often over water but also on land; 1–2 broods; 3–5 eggs; Mar–Jul.
FEEDING Fish; also frogs, insects, and spiders; stands quietly on the shore or in shallow water and strikes quickly; wades less often than larger herons.
HABITAT Swampy thickets, or dry land close to water; coastal wetlands in winter.
LENGTH 14½–15½ in (37–39cm)
WINGSPAN 25–27in (63–68cm)

Great Blue Heron

Ardea herodias

This is one of the three largest herons in the world—the Great Blue in North America, the Gray in Eurasia, and the Cocoi in South America—all of which are all interrelated, but classified separately. The Great Blue Heron is a common inhabitant of a variety of North American waterbodies, from marshes to swamps, as well as along sea coasts. As opportunistic feeders, they sometimes kill and swallow ducklings and small mammals.

dark wing tips

brownish body

dark tail

crooked neck

gray neck

white face

dark bill

blue-gray body

shaggy plumes

dark legs

VOICE *Mostly silent; gives a loud, barking* squawk *or* crank *in breeding colonies or when disturbed.*
NESTING *Nest of twigs and branches; usually in colonies, but also singly; in trees, over water or ground; 2–4 eggs; 1–2 broods; Feb–Aug.*
FEEDING *Catches prey with quick jab of bill; primarily fish and frogs.*
HABITAT *Wetlands, such as marshes, lake edges, rivers, and swamps; marine habitats such as tidal grass flats.*
LENGTH 2¾–4¼ft (0.9–1.3m)
WINGSPAN 5¼–6½ft (1.6–2m)

Great Egret

Ardea alba

Unlike other egrets, the Great Egret apparently prefers to forage alone; it maintains space around itself, defending a territory of 10ft (3m) in diameter from other wading birds. This territory "moves" with the bird as it feeds. In years of scarce food supplies, a nestling may kill a sibling, permitting the survival of at least one bird.

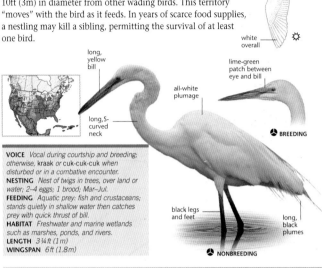

large size

white overall

long, yellow bill

lime-green patch between eye and bill

all-white plumage

long, S-curved neck

BREEDING

black legs and feet

long, black plumes

NONBREEDING

VOICE *Vocal during courtship and breeding; otherwise,* kraak *or* cuk-cuk-cuk *when disturbed or in a combative encounter.*
NESTING *Nest of twigs in trees, over land or water; 2–4 eggs; 1 brood; Mar–Jul.*
FEEDING *Aquatic prey: fish and crustaceans; stands quietly in shallow water then catches prey with quick thrust of bill.*
HABITAT *Freshwater and marine wetlands such as marshes, ponds, and rivers.*
LENGTH *3¼ft (1m)*
WINGSPAN *6ft (1.8m)*

Cattle Egret

Bubulcus ibis

Unlike most other herons, the Cattle Egret is a grassland species that rarely wades in water, and is often seen with livestock, feeding on the insects disturbed by their feet. It was first seen in Florida in 1941, but expanded rapidly and has now bred in over 40 US states and up into the southern provinces of Canada.

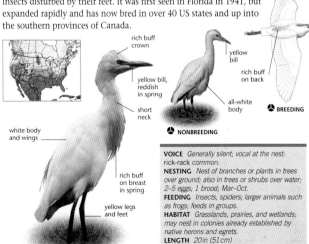

rich buff crown

yellow bill, reddish in spring

short neck

white body and wings

rich buff on breast in spring

yellow legs and feet

BREEDING

yellow bill

rich buff on back

all-white body

BREEDING

NONBREEDING

VOICE *Generally silent; vocal at the nest:* rick-rack *common.*
NESTING *Nest of branches or plants in trees over ground; also in trees or shrubs over water; 2–5 eggs; 1 brood; Mar–Oct.*
FEEDING *Insects, spiders; larger animals such as frogs; feeds in groups.*
HABITAT *Grasslands, prairies, and wetlands; may nest in colonies already established by native herons and egrets.*
LENGTH *20in (51cm)*
WINGSPAN *31in (78cm)*

Pelicans
& Relatives

Pelicans and their relatives are huge fish-eating birds with four toes connected by leathery webs, and with fleshy, elastic pouches beneath their bills.

Pelicans are buoyant swimmers and excellent fliers, with long, broad wings. Flocks can be seen soaring to great heights on migration and when flying to feeding grounds. They feed by sweeping with open bills for fish, often cooperatively, or by plunging from a height to scoop up fish and water in their large, flexible bill pouches.

Cormorants are medium to large waterbirds, some marine, others freshwater, with broad, long wings, rounded tails, short, strong legs and hook-tipped bills. When hunting for fish, cormorants dive from the surface of the water and then swim underwater with closed wings, using their webbed toes for propulsion. Most cormorants nest on cliff ledges, although some prefer trees.

LARGE COLONIES
The American White Pelican is highly social and is often seen feeding and roosting in large groups.

Northern Gannet

Morus bassanus

The Northern Gannet is known for its spectacular headfirst dives during frantic, voracious foraging in flocks of hundreds to thousands for surface-schooling fish. This colonially nesting bird breeds in just six locations in northeastern Canada. The Northern Gannet was the first species to have its total world population estimated, at 83,000 birds in 1939. Numbers have since increased.

black wing tip

long, pointed wing

yellow-orange nape

black-and-white mottled upperparts

yellow tinge to back of head

light blue eye

pointed gray bill

2ND YEAR

white upperparts

white underparts

pointed tail

VOICE *Loud landing call arrrr, arrah, or urrah rah rah; hollow groan oh-ah during take-off;* **krok** *call at sea.*
NESTING *Large pile of mud, seaweed, and rubbish, glued with guano, on bare rock or soil; 1 egg; 1 brood; Apr–Nov.*
FEEDING *Plunge-dives headfirst into water and often swims with its wings underwater to catch mackerel, herring, capelin, and cod.*
HABITAT *Isolated rock stacks on small islands or cliffs while breeding; continental shelf waters in winter.*
LENGTH *2¾–3½ ft (0.8–1.1m)*
WINGSPAN *5½ft (1.7m)*

American White Pelican ⓢ

Pelecanus erythrorhynchos

This enormous bird, with its distinctive, oversized bill, is a social inhabitant of large lakes and marshes in western North America. Most of its population is concentrated in a handful of large colonies in isolated wetlands in deserts and prairies. The American White Pelican forms cooperative foraging flocks, which beat their wings to drive fish into shallow water, where they can be caught more easily.

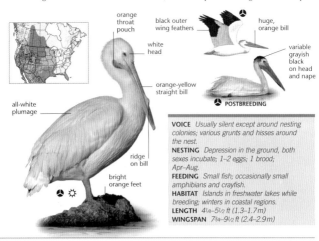

orange throat pouch

black outer wing feathers

huge, orange bill

white head

variable grayish black on head and nape

orange-yellow straight bill

POSTBREEDING

all-white plumage

ridge on bill

bright orange feet

VOICE *Usually silent except around nesting colonies; various grunts and hisses around the nest.*
NESTING *Depression in the ground, both sexes incubate; 1–2 eggs; 1 brood; Apr–Aug.*
FEEDING *Small fish; occasionally small amphibians and crayfish.*
HABITAT *Islands in freshwater lakes while breeding; winters in coastal regions.*
LENGTH *4¼–5½ ft (1.3–1.7 m)*
WINGSPAN *7¾–9½ ft (2.4–2.9 m)*

Brandt's Cormorant ⓢ

Phalacrocorax penicillatus

Brandt's Cormorant is the only cormorant with a blue chin, edged with a pale brownish patch at its lower end. Unlike the Double-crested Cormorant, most Brandt's fly with their necks straight. It depends heavily on food from the nutrient-rich California Current along the West Coast during breeding season.

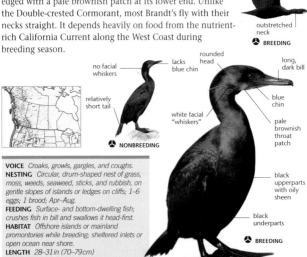

black overall

outstretched neck

BREEDING

rounded head

no facial whiskers

lacks blue chin

long, dark bill

relatively short tail

blue chin

white facial "whiskers"

pale brownish throat patch

NONBREEDING

black upperparts with oily sheen

black underparts

BREEDING

VOICE *Croaks, growls, gargles, and coughs.*
NESTING *Circular, drum-shaped nest of grass, moss, weeds, seaweed, sticks, and rubbish, on gentle slopes of islands or ledges on cliffs; 1–6 eggs; 1 brood; Apr–Aug.*
FEEDING *Surface- and bottom-dwelling fish; crushes fish in bill and swallows it head-first.*
HABITAT *Offshore islands or mainland promontories while breeding; sheltered inlets or open ocean near shore.*
LENGTH *28–31 in (70–79 cm)*
WINGSPAN *3½ ft (1.1 m)*

Double-crested Cormorant

Phalacrocorax auritus

This species is the most widespread of the North American cormorants. It often flies high over land in V-shaped flocks, but is mostly seen swimming low in the water with its head and neck visible, or resting on trees and rocks, sometimes with its wings spread to dry and drain their flight feathers after fishing bouts. Cormorants gain ballast for diving by temporarily filling their hollow primaries with water like submarines.

no crest

browner plumage overall

NONBREEDING

BREEDING

long neck

white crest

bluish eye

orange facial skin

black overall

black underparts

P. a. cincinatus
WESTERN; BREEDING

VOICE *Deep gruntlike calls; t-t-t-t call before taking off, and urg-urg-urg before landing; prolonged arr-r-r-r-t-t while mating, and eh-hr as threat.*
NESTING *Nests of sticks, seaweed, and trash, lined with grass; on ground or in trees usually in colonies; 3-5 eggs; 1 brood; Apr–Aug.*
FEEDING *Slow-moving or schooling fish; insects, crustaceans, amphibians, and, rarely, voles and snakes.*
HABITAT *Aquatic habitats while breeding, including ponds, lakes, rivers, seashores and marinas; coastlines and sandbars in winter.*
LENGTH 28–35in (70–90cm)
WINGSPAN 3½–4ft (1.1–1.2m)

Great Cormorant

S

Phalacrocorax carbo

The Great Cormorant is the most widely distributed cormorant species in the world. It sometimes breeds in mixed colonies with Double-crested Cormorants. They look similar from a distance, however Great Cormorants can be distinguished by their stouter bill, larger size, and white throat when breeding. The Great Cormorant can dive to depths of 115ft (35m) to catch prey.

outstretched head

large head with flat forehead

thick bill with hooked tip

neck kinked in flight

orange-yellow patch of skin near bill

brown neck

white throat

mostly white underparts

long, black neck

glossy black underparts with greenish scalloping

long body with glossy black upperparts

short, black legs and webbed feet

long, broad tail

🦅 BREEDING

VOICE *Deep, guttural calls at nesting and roosting site; otherwise silent.*
NESTING *Mound of seaweed, sticks, and debris, on cliff ledges and flat rocks above high-water mark; 3–5 eggs; 1 brood; Apr–Aug.*
FEEDING *Fish and small crustaceans; smaller prey swallowed underwater, while larger prey brought to surface.*
HABITAT *Cliff edges on rocky coasts while breeding; shallow coastal waters in winter.*
LENGTH *33–35in (84–90cm)*
WINGSPAN *4¼–5¼ft (1.3–1.6m)*

Pelagic Cormorant

S

Phalacrocorax pelagicus

The Pelagic Cormorant is the smallest cormorant species in North America. More solitary than other North American species, this bird is most visible at its roosting sites, where it spends much of its time drying its feathers. It is threatened by the disturbance of its nesting colonies, oil spills, entanglement in fishing nets, and pollution.

outstretched head and tail level in flight

small head

tufts on crown and nape

red patch at base of bill

all-dark face

thin, dark bill with blunt or hooked end

brownish bronze upperparts

thin, pale bill

blackish breast and belly

long, thin neck with white flecks

glossy purple tinge on neck

glossy green to greenish bronze on upperparts

iridescent greenish black underparts

white patch on flank

long, blackish tail

🦅 BREEDING

VOICE *Females: igh-ugh, similar to ticking grandfather clock; males: purring or arr-arr-arr; both utter croaks, hisses, and low groans.*
NESTING *Saucer-shaped nest of grass, sticks, feathers, and marine debris, cemented to cliff face with guano; 3–5 eggs; 1 brood; May–Oct.*
FEEDING *Medium-sized fish; also invertebrates (shrimps, worms, and hermit crabs).*
HABITAT *Rocky coasts, bays, harbours, lagoons; steep cliffs on islands while breeding.*
LENGTH *20–30in (51–76cm)*
WINGSPAN *3¼–4ft (1–1.2m)*

Birds of Prey

Only one species of vulture occurs in Canada: the Turkey Vulture, whose acute sense of smell enables it to detect carrion hidden from sight beneath the forest canopy. This vulture can stay in the air for hours on end, using lift provided by updrafts.

Falcons range in size from the diminutive American Kestrel to the large, powerful Gyrfalcon. They also include the Merlin, the Prairie Falcon, and the fast-diving Peregrine Falcon. Falcon prey ranges from insects to large hare-sized mammals and birds.

Eagle and hawk species cover a wide range of raptors of varying sizes and hunting methods. Forest-dwelling hawks rely on speed and stealth to pounce on small birds among the trees, while the Osprey hovers over water until it sees a fish below, then dives to pluck its prey out of the water with its talons.

DOUBLE SHOT
When there are lots of fish running in a tight school, the Osprey has the strength and skill to catch two with one dive.

Turkey Vulture

Cathartes aura

The most widely distributed vulture in North America, the Turkey Vulture is found in most of the US and has expanded its range into southern Canada. It possesses a better sense of smell than the Black Vulture, which often follows it and displaces it from carcasses. The Turkey Vulture's habit of defecating down its legs may serve to cool it or to kill bacteria with its ammonia content.

long wings

brownish gray head

blackish back feathers, edged brown

silvery gray flight feathers

naked skin

small, red head

brownish back

black underparts

pink legs

long tail

VOICE *Silent, but will hiss at intruders; also grunts.*
NESTING *Dark recesses under large rocks or stumps, on rocky ledges in caves, in mammal burrows, and abandoned buildings; 1-3 eggs; 1 brood; Mar.–Aug.*
FEEDING *Wild and domestic carrion: mammals, birds, reptiles, amphibians, fish; occasionally live prey (nestlings or trapped birds).*
HABITAT *Mixed farmland, forests, and urban areas; forested hillsides and abandoned structures while nesting.*
LENGTH *25–32in (64–81cm)*
WINGSPAN *5½–6ft (1.7–1.8m)*

American Kestrel

S

Falco sparverius

The smallest of the North American falcons, the American Kestrel features typically pointed wings, a "tooth and notch" bill structure, and the dark brown eyes typical of falcons, but shorter toes. They often hover over fields to capture insects and small mammals. Male and female American Kestrels show striking differences in plumage, the former also being larger.

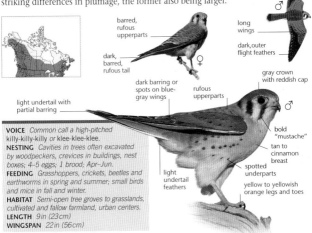

barred, rufous upperparts

dark, barred, rufous tail

♀

♂

long wings

dark, outer flight feathers

dark barring or spots on blue-gray wings

rufous upperparts

gray crown with reddish cap

light undertail with partial barring

♂

bold "mustache"

tan to cinnamon breast

light undertail feathers

spotted underparts

yellow to yellowish orange legs and toes

VOICE *Common call a high-pitched killy-killy-killy or klee-klee-klee.*
NESTING *Cavities in trees often excavated by woodpeckers, crevices in buildings, nest boxes; 4–5 eggs; 1 brood; Apr–Jun.*
FEEDING *Grasshoppers, crickets, beetles and earthworms in spring and summer; small birds and mice in fall and winter.*
HABITAT *Semi-open tree groves to grasslands, cultivated and fallow farmland, urban centers.*
LENGTH *9in (23cm)*
WINGSPAN *22in (56cm)*

Merlin

S

Falco columbarius

Merlins are small, fast-flying falcons that can overtake and capture a wide variety of prey. They can turn on a dime, and use their long, thin toes, typical of falcons, to pluck small birds from the air. Males are smaller than females, and different in color. Both males and females show geographical color variations.

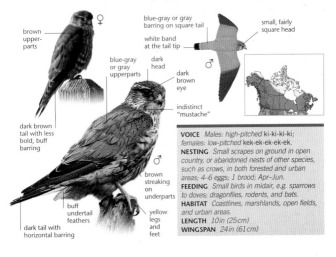

♀

brown upperparts

blue-gray or gray barring on square tail

white band at the tail tip

blue-gray or gray upperparts

dark head

♂

dark brown eye

small, fairly square head

indistinct "mustache"

dark brown tail with less bold, buff barring

♂

brown streaking on underparts

buff undertail feathers

yellow legs and feet

dark tail with horizontal barring

VOICE *Males: high-pitched ki-ki-ki-ki; females: low-pitched kek-ek-ek-ek-ek.*
NESTING *Small scrapes on ground in open country, or abandoned nests of other species, such as crows, in both forested and urban areas; 4–6 eggs; 1 brood; Apr–Jun.*
FEEDING *Small birds in midair, e.g. sparrows to doves; dragonflies, rodents, and bats.*
HABITAT *Coastlines, marshlands, open fields, and urban areas.*
LENGTH *10in (25cm)*
WINGSPAN *24in (61cm)*

Gyrfalcon

(S)

Falco rusticolus

The Arctic-breeding Gyrfalcon is used to harsh environments. It is the largest of all the falcons and is also the official bird of the Northwest Territories. It uses its speed to pursue prey in a "tail chase," sometimes striking its quarry on the ground, but also in flight. Three forms are known, ranging from almost pure white to gray and dark.

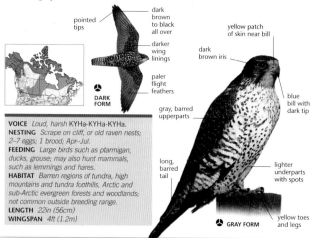

pointed tips

dark brown to black all over

darker wing linings

paler flight feathers

DARK FORM

yellow patch of skin near bill

dark brown iris

gray, barred upperparts

blue bill with dark tip

long, barred tail

lighter underparts with spots

GRAY FORM

yellow toes and legs

VOICE *Loud, harsh KYHa-KYHa-KYHa.*
NESTING *Scrape on cliff, or old raven nests; 2–7 eggs; 1 brood; Apr–Jul.*
FEEDING *Large birds such as ptarmigan, ducks, grouse; may also hunt mammals, such as lemmings and hares.*
HABITAT *Barren regions of tundra, high mountains and tundra foothills, Arctic and sub-Arctic evergreen forests and woodlands; not common outside breeding range.*
LENGTH *22in (56cm)*
WINGSPAN *4ft (1.2m)*

Prairie Falcon

(S)

Falco mexicanus

Prairie Falcons live in the arid regions of North America and blend in well with their surroundings (cliff faces and dry grass). They chase their prey close to the ground and do not often dive on prey from a great height. In some areas, breeding is linked to the emergence of ground squirrels, an important prey. The underwing pattern with blackish feathers in the "wingpits" is distinctive to this species.

yellow patch of skin near bill

yellow eye-ring

light head and "mustache"

long, pointed wings

longish tail

white cheek

light, sandy brown upperparts with incomplete barring

light underparts with brown spots

distinctive, triangle-shaped patch on wingpit feathers

VOICE *Repeated shrill kik-kik-kik-kik-kik.*
NESTING *Slight, shallow scrapes, sometimes in a cavity on high cliff ledges or bluffs; often in company with Common Ravens, Golden Eagles, and Red-tailed Hawks; 3–6 eggs; 1 brood; Mar–Jul.*
FEEDING *Small to medium-sized birds and small mammals, such as ground squirrels.*
HABITAT *Open plains, prairies and grasslands, dotted with buttes or cliffs.*
LENGTH *16in (41cm)*
WINGSPAN *3¼ft (1m)*

yellow legs and toes

light undertail feathers

Peregrine Falcon

Falco peregrinus

Peregrine Falcons are distributed worldwide and are long-distance travelers—"Peregrine" means "wanderer." It can dive from great heights at speeds of up to 200mph (320kmph)—a technique known as "stooping." Like all true falcons, this species has a pointed "tooth" on its upper beak and a "notch" on the lower one, apparently used to sever the neck vertebrae of captured prey to kill it. Its breeding ability was drastically reduced in the 1950s to 1980s due to insecticide poisoning, but with conservation efforts the population has rebounded.

long, pointed wings

short tail

dark "hood" on head

yellow eye-ring

dark spots on light buff breast

bluish gray upperparts

light underparts with horizontal barring

brown upperparts

streaked underparts

light yellow or bluish gray legs and toes

yellow toes and legs

VOICE *Sharp* cack-cack-cack *when alarmed.*
NESTING *Shallow scrape on cliff or building (nest sites are used year after year); 2–5 eggs; 1 brood; Mar–Jun.*
FEEDING *Dives on prey—birds of various sizes in flight; also pigeons in cities, bats; occasionally steals prey, including fish and rodents, from other raptors.*
HABITAT *Cliffs along coasts, inland mountain ranges, scrubland and salt marshes, cities with tall buildings.*
LENGTH *16in (41cm)*
WINGSPAN *3¼–3½ft (1–1.1m)*

Osprey

(S)

Pandion haliaetus

Sometimes referred to as the "fish hawk" or "fish eagle," the Osprey is the only bird of prey in North America that feeds almost exclusively on live fish. Sharp spicules (tiny, spike-like growths) on the pads of its feet, reversible outer toes, and an ability to lock its talons in place enable it to hold onto slippery fish. The Osprey is the provincial bird of Nova Scotia.

dark band running across wing

wing tips at slight backward angle

dark brown upperparts

wings bowed while soaring

barred tail

finely barred underwings

black eye stripe

crest on head

black mask on face

speckled chest

black bill

white underparts

pale gray legs and feet

VOICE *Slow, whistled notes, falling in pitch: tiooop, tioooop, tiooop; also screams by displaying male.*
NESTING *Twig nest on tree, cliff, rock pinnacles, ground, hydro towers, channel markers; 1–4 eggs; 1 brood; Mar–Aug.*
FEEDING *Dives to catch fish up to top 3ft (90cm) of water.*
HABITAT *Forests near water, e.g. lakes, rivers, coastal waters, estuaries.*
LENGTH *21–23in (53–58cm)*
WINGSPAN *5–6ft (1.5–1.8m)*

Bald Eagle

(S)

Haliaeetus leucocephalus

The Bald Eagle, although an opportunist, prefers to scavenge on carrion and steal prey from other birds, including Ospreys. The use of DDT causing reproductive failure, as well as bounties, led to it being declared endangered in 1967. Now frequenting landfill sites, its numbers have since rebounded, especially on east and west coasts.

white head

brown body

white tail

yellow, hooked bill

dark brown overall

white tail

dark bill starting to turn yellow at base

pure white head with yellow eyes

dark chocolate-brown overall

long, wedge-shaped, white tail

yellow legs and feet

🕐 **1ST YEAR**

VOICE *Surprisingly high-pitched voice, 3–4 notes followed by a rapidly descending series.*
NESTING *Huge stick nest, usually in tallest tree; 1–3 eggs; 1 brood; Mar–Sep.*
FEEDING *Carrion, especially fish; birds, mammals; steals fish from Osprey.*
HABITAT *Forested areas near water while breeding; along major rivers and coastal areas in winter.*
LENGTH *28–38in (71–96cm)*
WINGSPAN *6½ft (2m)*

Northern Harrier

Circus cyaneus

The Northern Harrier is most often seen flying low in search of food. A white rump, V-shaped wings, and tilting flight make this species easily identifiable. The blue-gray males are strikingly different from the dark-brown females. The bird has an owl-like face, which contains stiff feathers to help channel in sounds from prey.

black wing tips

white ring around face ♀

brown upperparts

white rump

bluish gray head

dark bill with yellow skin near bluish base

bluish gray upperparts

♂

white underparts with reddish brown markings

gray uppertail with light undertail feathers

VOICE Kek *in rapid succession at nest, becoming more high-pitched in alarm.*
NESTING *Platform of sticks on ground in open, wet field; 4–6 eggs; 1 brood; Apr–Sep.*
FEEDING *Rodents like mice and muskrats; also birds, frogs, reptiles; occasionally larger prey such as rabbits.*
HABITAT *Open wetlands while breeding; open habitats such as deserts, coastal sand dunes, and grasslands in winter.*
LENGTH *18–20in (46–51cm)*
WINGSPAN *3½–4ft (1.1m–1.2m)*

Sharp-shinned Hawk

Accipiter striatus

This small, swift hawk is quite adept at capturing songbirds in flight, occasionally even taking species larger than itself. The Sharp-shinned Hawk's short, rounded wings and long tail allow it to make abrupt turns and fast dashes in thick woods and dense shrubby terrain. Its prey is plucked before being consumed or fed to the nestlings.

grayish blue upperparts ♂

yellow legs and toes

slightly browner upperparts than male

grayish blue crown

reddish yellow eye

wide, dark, horizontal bars on gray tail

reddish brown bars on underparts

VOICE *High-pitched, repeated* kiu kiu kiu; *squealing sound when disturbed at nest.*
NESTING *Sturdy nest of sticks lined with twigs or pieces of bark; 3–4 eggs; 1 brood; Mar–Jun.*
FEEDING *Small birds, such as sparrows and wood-warblers, caught either on the wing or while perched.*
HABITAT *Coniferous forests and mixed hardwood-conifer woodlands.*
LENGTH *11in (28cm)*
WINGSPAN *23in (58cm)*

white, fluffy undertail feathers ♀

Cooper's Hawk

Accipiter cooperii

A secretive bird, the Cooper's Hawk is capable of quickly maneuvering through dense vegetation. They are now numerous in urban areas, mainly taking pigeons, doves, and birds at feeders. Should a human approach its nest, the brooding adult will quietly glide down and away from the nest tree rather than attack the intruder.

grayish blue overall

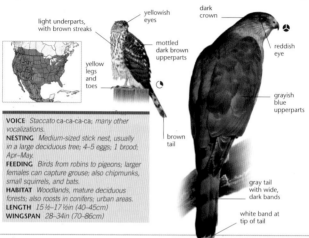

light underparts, with brown streaks

yellowish eyes

mottled dark brown upperparts

yellow legs and toes

dark crown

reddish eye

grayish blue upperparts

brown tail

gray tail with wide, dark bands

white band at tip of tail

VOICE *Staccato ca-ca-ca-ca; many other vocalizations.*
NESTING *Medium-sized stick nest, usually in a large deciduous tree; 4–5 eggs; 1 brood; Apr.–May.*
FEEDING *Birds from robins to pigeons; larger females can capture grouse; also chipmunks, small squirrels, and bats.*
HABITAT *Woodlands, mature deciduous forests; also roosts in conifers; urban areas.*
LENGTH *15½–17½in (40–45cm)*
WINGSPAN *28–34in (70–86cm)*

Northern Goshawk

Accipiter gentilis

The Northern Goshawk is secretive by nature and not easily observed, even where it is common. Often nesting in mature tree farms, it is a fierce and noisy defender of its territories, nests, and young and will attack even humans. Spring hikers and horse riders occasionally discover Northern Goshawks by wandering into their territory only to be driven off by the angry occupants.

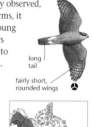

long tail

fairly short, rounded wings

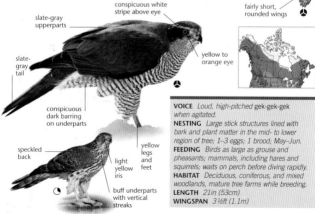

slate-gray upperparts

slate-gray tail

conspicuous white stripe above eye

yellow to orange eye

conspicuous dark barring on underparts

speckled back

light yellow iris

yellow legs and feet

buff underparts with vertical streaks

VOICE *Loud, high-pitched gek-gek-gek when agitated.*
NESTING *Large stick structures lined with bark and plant matter in the mid- to lower region of tree; 1–3 eggs; 1 brood; May–Jun.*
FEEDING *Birds as large as grouse and pheasants; mammals, including hares and squirrels; waits on perch before diving rapidly.*
HABITAT *Deciduous, coniferous, and mixed woodlands, mature tree farms while breeding.*
LENGTH *21in (53cm)*
WINGSPAN *3½ft (1.1m)*

Red-shouldered Hawk ⓢ

Buteo lineatus

The Red-shouldered Hawk has widespread populations in the East and northeast, and in the Midwest Great Plains and US West Coast. Eastern birds are divided into four subspecies; western populations belong to the subspecies *B. l. elegans*. The red shoulder patches are not always evident, but the striped tail and translucent "windows" in the wings are easily identifiable.

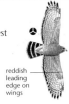

reddish leading edge on wings

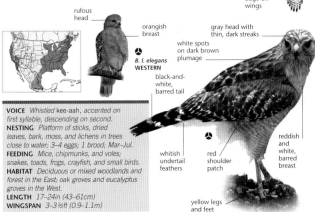

rufous head

orangish breast

B. l. elegans WESTERN

black-and-white, barred tail

gray head with thin, dark streaks

white spots on dark brown plumage

reddish and white, barred breast

whitish undertail feathers

red shoulder patch

yellow legs and feet

VOICE Whistled *kee-aah*, accented on first syllable, descending on second.
NESTING Platform of sticks, dried leaves, bark, moss, and lichens in trees close to water; 3–4 eggs; 1 brood; Mar–Jul.
FEEDING Mice, chipmunks, and voles; snakes, toads, frogs, crayfish, and small birds.
HABITAT Deciduous or mixed woodlands and forest in the East; oak groves and eucalyptus groves in the West.
LENGTH 17–24in (43–61cm)
WINGSPAN 3–3½ft (0.9–1.1m)

Broad-winged Hawk ⓢ

Buteo platypterus

Broad-winged Hawks migrate in huge flocks or "kettles," soaring on rising thermals. The majority log more than 4,000 miles (6,500km) before ending up in Brazil, Bolivia, and some of the Caribbean islands. Adults are easily identified by a broad, white-and-black band on their tails. Broad-winged Hawks have two color forms, the light one being more common than the dark, sooty brown one.

one to two broad, white bands visible on tail

dark border on edges of wings

indistinct "mustache"

upperparts brown with white flecking

pale underparts, with conspicuous, tear-shaped, brown spots

short, yellow feet

VOICE High-pitched *peeoweee*, first note shorter and higher-pitched.
NESTING Platform of fresh twigs or dead sticks, often on old squirrel, hawk, or crow nest in tree; 2–3 eggs; 1 brood; Apr–Aug.
FEEDING Small mammals, toads, frogs, snakes, grouse chicks, insects, and spiders; crabs in winter.
HABITAT Habitats with deciduous, conifers, and mixed trees, with clearings and water.
LENGTH 13–17in (33–43cm)
WINGSPAN 32–39in (81–100cm)

Red-tailed Hawk

Buteo jamaicensis

The Red-tailed Hawk is the most widely distributed hawk in North America. As many as 15 subspecies have been described to date, varying in coloration, tail markings, and size. The all-dark Harlan's Hawk, which breeds in Alaska and northwestern Canada, is considered by some to be a separate species and by others as a subspecies of the Red-tailed hawk. While it occasionally stoops on prey, the Red-tailed Hawk usually adopts a sit-and-wait approach.

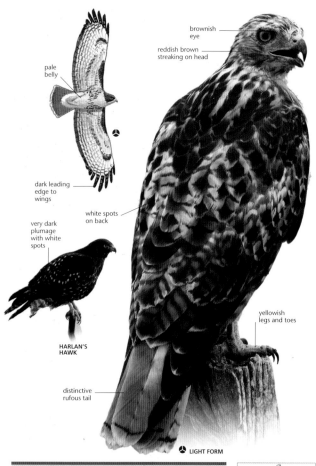

brownish eye

reddish brown streaking on head

pale belly

dark leading edge to wings

white spots on back

very dark plumage with white spots

HARLAN'S HAWK

distinctive rufous tail

yellowish legs and toes

LIGHT FORM

VOICE *Kee-eee-arrr that rises then descends over a period of 2–3 seconds.*
NESTING *Large platform of sticks, twigs on top of tall tree, cliff, building, or billboard; 2 eggs; 1 brood; Feb–Sep.*
FEEDING *Small mammals, such as mice, rats; birds including pheasant, quail; small reptiles; carrion.*
HABITAT *Open areas in scrub desert, grasslands, agricultural fields, and woodlands; occasionally in urban areas.*
LENGTH *18–26in (46–65cm)*
WINGSPAN *3½–4¼ft (1.1–1.3m)*

Swainson's Hawk

Buteo swainsoni

Swainson's Hawk is famous for its spectacular 6,000-mile (9,650km) fall migration from the Canadian prairies to the lower regions of South America, when thousands can be observed soaring in the air at any one time. While migrating, this hawk averages 125 miles (200km) a day. There are three color forms: light, dark, and an intermediate form between the two.

long pointed wings

dark chest

LIGHT FORM

dark brown head and breast

spotted underparts

DARK FORM

white face and chin

slender shape overall

pale reddish upper chest

white underbelly

longish tail

wing tips reach end of tail when perched

LIGHT FORM

VOICE *Shrill, plaintive scream* kreeeee *in alarm; high-pitched* keeeoooo *fading at the end.*
NESTING *Bulky, flimsy pile of sticks or various debris, in solitary tree or on utility poles; 1–4 eggs; 1 brood; Apr–Jul.*
FEEDING *Ground squirrels, gophers, mice, voles, bats, rabbits; snakes, lizards, songbirds.*
HABITAT *Scattered trees along streams while breeding; areas of open woodland, grassland, and agricultural fields.*
LENGTH *19–22in (48–56cm)*
WINGSPAN *4½ft (1.4m)*

Ferruginous Hawk

Buteo regalis

This inhabitant of open country is the largest North American hawk. As a versatile nester, it builds its stick nests on cliffs or nearly level ground, trees, and manmade structures like farm buildings. Regrettably, its preference for prairie dogs, which are declining because of habitat loss, shooting, and pesticide use, threatens Ferruginous Hawk populations.

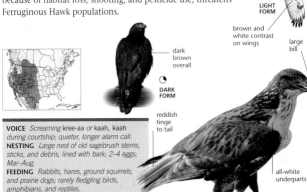

LIGHT FORM

brown and white contrast on wings

large bill

dark brown overall

DARK FORM

reddish tinge to tail

all-white underparts

fully feathered legs

LIGHT FORM

VOICE *Screaming* kree-aa *or* kaah, kaah *during courtship; quieter, longer alarm call.*
NESTING *Large nest of old sagebrush stems, sticks, and debris, lined with bark; 2–4 eggs; Mar–Aug.*
FEEDING *Rabbits, hares, ground squirrels, and prairie dogs; rarely fledgling birds, amphibians, and reptiles.*
HABITAT *Low-elevation grasslands with cliffs or isolated trees for nesting.*
LENGTH *22–27in (56–69cm)*
WINGSPAN *4¼–4½ft (1.3–1.4m)*

Rough-legged Hawk ⓢ

Buteo lagopus

The Rough-legged Hawk is known for its extensive variation in plumage, ranging from almost completely black to nearly cream or whitish. The year to year fluctuation in numbers of breeding pairs in a given region strongly suggest that this species is nomadic, moving about as a response to the availability of its prey.

dark tail band

bold black patch

dark wing tips

pale head

short, broad head

barred underparts

♂

thin bands near tail tip

black belly

white tail with faint black band at tip

VOICE *Breeding birds: loud, cat-like mewing or thin whistles, slurred downward in alarm.*
NESTING *Bulky mass of sticks, lined with grasses, sedges, feathers and fur from prey, on cliff ledge; 2–6 eggs; 1 brood; Apr–Aug.*
FEEDING *Lemmings and voles in spring and summer; mice and shrews in winter; birds, ground squirrels, and rabbits year-round.*
HABITAT *Rough, open country with cliffs while breeding; edges of forests, treeless tundra.*
LENGTH *19–20in (48–51cm)*
WINGSPAN *4¼–4½ft (1.3–1.4m)*

Golden Eagle ⓢ

Aquila chrysaetos

Perhaps the most formidable of all North American birds of prey, the Golden Eagle is found mostly in the western part of the continent. It defends large territories ranging from 8–12 square miles (20–30 square kilometers). Although it appears sluggish, it is swift and agile, and employs a variety of hunting techniques to catch specific prey.

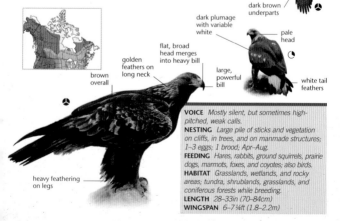

dark brown underparts

dark plumage with variable white

pale head

flat, broad head merges into heavy bill

golden feathers on long neck

large, powerful bill

white tail feathers

brown overall

heavy feathering on legs

VOICE *Mostly silent, but sometimes high-pitched, weak calls.*
NESTING *Large pile of sticks and vegetation on cliffs, in trees, and on manmade structures; 1–3 eggs; 1 brood; Apr–Aug.*
FEEDING *Hares, rabbits, ground squirrels, prairie dogs, marmots, foxes, and coyotes; also birds.*
HABITAT *Grasslands, wetlands, and rocky areas; tundra, shrublands, grasslands, and coniferous forests while breeding.*
LENGTH *28–33in (70–84cm)*
WINGSPAN *6–7¼ft (1.8–2.2m)*

Rails

The rail family is a diverse group of small to medium-sized marsh birds, chicken-like, with stubby tails and short, rounded wings. Rails and crakes inhabit dense marshland and are secretive and solitary, whereas coots and gallinules are seen on open water. Rallids look like weak fliers, but many migrate great distances at night. During nesting, members of each pair keep in close contact by calling loudly and clearly.

LIGHT ON ITS TOES
Common moorhens walk easily on floating vegetation thanks to their long toes.

Cranes

Cranes are large wading birds, superficially similar to storks and to the larger herons and egrets. Notable differences include long inner wing feathers forming a "bustle" on a standing crane and a straight neck while flying, instead of the tight S-curve seen in similar-sized herons. The Whooping Crane is the tallest bird in North America, standing nearly 5ft (1.5m) high.

CRANE RALLY
Large groups of Sandhill Cranes gather on feeding grounds in winter.

Yellow Rail

Ⓓ

Coturnicops noveboracensis

The secretive, nocturnal Yellow Rail is extremely difficult to observe
in its dense, grassy habitat, and is detected mainly by its voice.
It has a small head, almost no neck, a stubby bill, a plump,
almost tail-less body, and short legs. The bill of the male
turns yellow in the breeding season.

dangling legs

dark brown crown

stubby yellow to olive-gray bill

white patch on inner wing feathers

long tan stripes on blackish background

dark stripe runs from cheek to bill

buff or yellow breast

short tail

VOICE *Males: clicking calls, also descending
cackles, quiet croaking, and soft clucking.*
NESTING *Small cup of grasses and sedges,
on the ground or in a plant tuft above water,
concealed by overhanging vegetation;
8–10 eggs; 1 brood; May–Jun.*
FEEDING *Seeds, aquatic insects, small
crustaceans, freshwater snails.*
HABITAT *Brackish and freshwater marshes
and wet sedge meadows while breeding.*
LENGTH *7¼in (18.5cm)*
WINGSPAN *11in (28cm)*

King Rail

●

Rallus elegans

A scattered and localized breeder across eastern North America, the King Rail
depends on extensive freshwater marsh habitats with tall, emergent reeds and
cattails. Concealed by this vegetation, this chicken–like bird is rarely seen and
is most often detected by its distinctive calls.

rufous upperwing

brown stripe running down neck

reddish eye

heavy down-curved bill

boldly streaked upperparts

long, curved, yellow-orange bill

orangish breast

boldly barred, black-and-white flanks

short tail

VOICE *Males: low clacks or grunts; emits a
loud kik kik kik during breeding season.*
NESTING *Cup of vegetation, often hidden by
bent stems that form a canopy; 6–12 eggs;
2 broods; Feb–Aug.*
FEEDING *Insects, snails, spiders, and
crustaceans such as shrimps, crabs, and
barnacles; also fish, frogs, and seeds.*
HABITAT *Freshwater marshes while breeding;
coastal marshes in winter.*
LENGTH *15in (38cm)*
WINGSPAN *20in (51cm)*

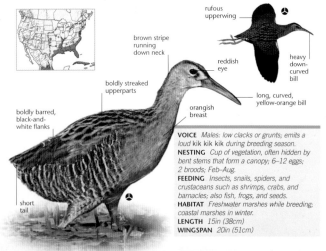

Virginia Rail

Rallus limicola

A smaller version of the King Rail, this freshwater marsh dweller is more often heard than seen. Distributed in a wide range, the Virginia Rail spends most of its time in thick, reedy vegetation, which it pushes using its "rail thin" body and flexible vertebrae. Although it spends most of its life walking, it can swim and dive to escape danger. It is a partial migrant that leaves its northern breeding grounds in the winter.

rufous upperwing

❧ BREEDING

dark outer wing feathers

diffused streaking

dark bill

dark, blotchy breast

❧ NONBREEDING

streaked black and brown upperparts

gray cheeks

curved, red bill

reddish brown breast

white undertail

❧ BREEDING

black-and-white barring on flanks

reddish legs and toes

VOICE *Series of pig-like grunting oinks, loud and sharp, then steadily softer; also emits a series of double notes ka-dik ka-dik.*
NESTING *Substantial cup of plant material, concealed by bent-over stems; 5–12 eggs; 1–2 broods; Apr–Jul.*
FEEDING *Stalks prey or waits and dives into water; snails, insects, and spiders, also seeds.*
HABITAT *Freshwater habitats while breeding; saltwater and freshwater marshes in winter.*
LENGTH 9½in (24cm)
WINGSPAN 13in (33cm)

Sora

S

Porzana carolina

The Sora is widely distributed but rarely seen. It breeds in freshwater marshes and migrates hundreds of miles south in winter despite its weak and hesitant flight. It swims well, with a characteristic head-bobbing action. The Sora can be spotted walking at the edge of emergent vegetation—its yellow bill and black mask distinguish it from other rails.

long, trailing legs

white markings on back

BREEDING

brown cheek patch

yellow bill

black mask

gray breast

yellowish green legs

short tail

BREEDING

reduced black on face

white barring on flanks

NONBREEDING

VOICE *Long, high, and loud, descending, horse-like whinny ko-wee-hee-hee-hee-hee; upslurred whistle.*
NESTING *Loosely woven basket of vegetation above water or in clumps of vegetation on water's surface; 8–11 eggs; 1 brood; May–Jun.*
FEEDING *Seeds of wetland plants, insects, spiders, and snails.*
HABITAT *Freshwater marshes with emergent vegetation while breeding.*
LENGTH *8½in (22cm)*
WINGSPAN *14in (36cm)*

Common Moorhen

S

Gallinula chloropus

The Common Moorhen is fairly widespread in southeastern Canada. At home on land and water, its long toes allow it to walk easily over floating vegetation and soft mud. When walking or swimming, the Common Moorhen jerks its short tail, revealing its white undertail feathers, and bobs its head.

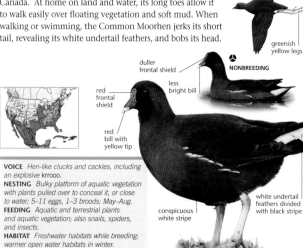

BREEDING

greenish yellow legs

NONBREEDING

duller frontal shield

less bright bill

red frontal shield

red bill with yellow tip

conspicuous white stripe

white undertail feathers divided with black stripe

BREEDING

VOICE *Hen-like clucks and cackles, including an explosive krrooo.*
NESTING *Bulky platform of aquatic vegetation with plants pulled over to conceal it, or close to water; 5–11 eggs, 1–3 broods; May–Aug.*
FEEDING *Aquatic and terrestrial plants and aquatic vegetation; also snails, spiders, and insects.*
HABITAT *Freshwater habitats while breeding; warmer open water habitats in winter.*
LENGTH *14in (36cm)*
WINGSPAN *21in (53cm)*

American Coot

Fulica americana

As the most abundant and widely distributed of North American rails, this duck-like bird commonly bobs its head forward and backward while swimming. Its lobed toes make it well adapted to swimming and diving, but are somewhat of an impediment on land. Its flight is clumsy; it becomes airborne with difficulty, running along the water surface before taking off. American Coots form large flocks on open water in winter, often associating with ducks—an unusual trait for a member of the rail family.

dull grayish plumage

white- edged feathers

white bill

🔺 BREEDING

black head

red eye

dark gray body

black ring on bill

🔺 BREEDING

lobed toes

long, greenish yellow legs

VOICE *Raucous clucks, grunts, and croaks and an explosive* **keek**.
NESTING *Bulky cup of plant material placed in aquatic vegetation on or near water; 5–15 eggs; 1–2 broods; Apr–Jul.*
FEEDING *Forages on or under shallow water and feeds on land; primarily herbivorous; also snails, insects, spiders, tadpoles, fish, and carrion.*
HABITAT *Open water habitats while breeding and in winter; ponds, marshes, reservoirs, lake edges, and saltwater inlets.*
LENGTH *15½in (40cm)*
WINGSPAN *24in (61cm)*

Sandhill Crane

Ⓢ

Grus canadensis

Sandhill Cranes are famous for their elaborate courtship dances, far-carrying vocalizations, and remarkable migrations. Their bodies are sometimes stained with a rusty color, supposedly because they probe into mud that contains iron and then preen, transferring color from bill to plumage. Sandhill Cranes are broadly grouped into "Lesser" and "Greater" populations that differ in the breeding ground locations and migration routes.

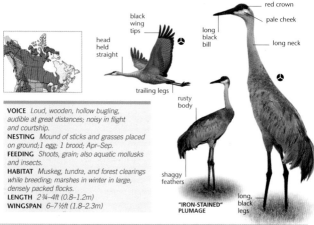

black wing tips

head held straight

trailing legs

red crown

pale cheek

long black bill

long neck

rusty body

shaggy feathers

"IRON-STAINED" PLUMAGE

long, black legs

VOICE *Loud, wooden, hollow bugling, audible at great distances; noisy in flight and courtship.*
NESTING *Mound of sticks and grasses placed on ground;1 egg; 1 brood; Apr–Sep.*
FEEDING *Shoots, grain; also aquatic mollusks and insects.*
HABITAT *Muskeg, tundra, and forest clearings while breeding; marshes in winter in large, densely packed flocks.*
LENGTH *2¾–4ft (0.8–1.2m)*
WINGSPAN *6–7½ft (1.8–2.3m)*

Whooping Crane

●

Grus americana

The Whooping Crane, although still endangered, has rebounded from near extinction due to an ambitious campaign of captive breeding and release in Canada and the US. They are territorial in both summer and winter, living in family groups. Whooping Cranes have an elaborate courtship display with leaps, wing flaps, and flinging of light objects such as grass and feathers.

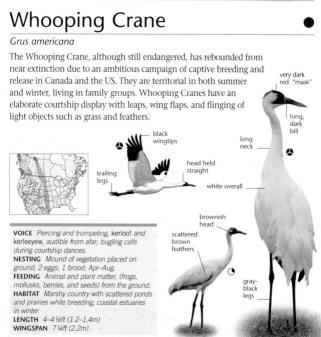

black wingtips

trailing legs

head held straight

long neck

very dark red "mask"

long, dark bill

white overall

brownish head

scattered brown feathers

gray-black legs

VOICE *Piercing and trumpeting, kerloo! and kerleeyew, audible from afar; bugling calls during courtship dances.*
NESTING *Mound of vegetation placed on ground; 2 eggs; 1 brood; Apr–Aug.*
FEEDING *Animal and plant matter, (frogs, mollusks, berries, and seeds) from the ground.*
HABITAT *Marshy country with scattered ponds and prairies while breeding; coastal estuaries in winter.*
LENGTH *4–4½ft (1.2–1.4m)*
WINGSPAN *7¼ft (2.2m)*

Shorebirds, Gulls, & Auks

Shorebirds include oystercatchers, avocets and stilts, plovers, sandpipers, and phalaropes. Many have long legs in proportion to their bodies, and a variety of bills, ranging from short to long, thin, thick, straight, down-curved and up-curved.

Canadian gull species all share a similar stout body shape, sturdy bills and webbed toes. Nearly all are scavengers and are often found near coastal areas, including fishing ports, or inland, especially in urban areas and garbage dumps.

Terns are specialized, long-billed predators that dive for fish. More slender and elegant than gulls, nearly all are immediately recognizable when breeding, due to their black caps and long, pointed bills.

Auks, murres, and puffins come to land only to breed. Most nest in colonies on sheer cliffs overlooking the ocean, but puffins excavate burrows in the ground, and some murrelets nest away from predators high up in treetops far inland.

ON THE MOVE
Dunlins and other sandpipers gather in large, highly coordinated flocks on migration.

Black Oystercatcher

Haematopus bachmani

This large, striking oystercatcher is instantly obvious because of its all-dark plumage and contrasting pale eyes and colorful bill. It is restricted to rocky coasts, where it feeds in pairs or family groups, using well-defined territories in summer. In winter, the birds gather in larger flocks, sometimes numbering in the hundreds, where mussels are abundant.

long, orange-red bill

broad, powerful wings

dull orange eye-ring

dark tip of bill

dark eye

orange-red eye-ring

bright yellow eye

thick, pink legs

dark brown to black body

VOICE *Loud, whistled* wheeu *in flight; alarm call sharper* wheep; *series of whistles in courtship.*
NESTING *Simple scrape above high-tide line, lined with broken shells and pebbles; 1–3 eggs; 1 brood; May–Jun.*
FEEDING *Mollusks (mussels, limpets); crustaceans (crabs, barnacles); rarely oysters.*
HABITAT *Rocky headlands or beaches while breeding; rocky jetties in winter.*
LENGTH *16½–18½in (42–47cm)*
WINGSPAN *30–34in (77–86cm)*

Black-necked Stilt

Himantopus mexicanus

This tall, slender shorebird is a familiar sight at ponds and lagoons in the southern Canadian prairies. It is remarkably long-legged: in flight, it often crosses its trailing feet as if for extra support. Breeding takes place in small colonies. In winter, small flocks of about 25 individuals feed quietly in sheltered areas or aggressively drive visitors away with raucous calls.

no white spot above red eye

long, slender neck

long, angular, black wings

black upperparts

white spot above red eye

black mask encircles eye

slender, tapered body

white underparts

long, needle-like black bill

♂

long, bright pink legs

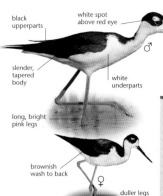

brownish wash to back

duller legs than male

♀

VOICE *Loud, continuous poodle-like* yip-yip-yip, *in flight or when alarmed.*
NESTING *Simple scrape lined with grass in soft soil; 4 eggs; 1 brood; Apr–May.*
FEEDING *Picks food off surface of shallow water; tadpoles, shrimps, snails, flies, worms, clams, small fish, and frogs.*
HABITAT *Marshes, shallow ponds, lake margins, and reservoirs while breeding and in winter.*
LENGTH *14–15½in (35–39cm)*
WINGSPAN *29–32in (73–81cm)*

American Avocet

Recurvirostra americana

With its long, thin, and upturned bill, this graceful, long-legged shorebird is unmistakable when foraging. When it takes off, its striking plumage pattern is clearly visible. It is the only one of the four avocet species in the world that changes plumage when breeding. Breeding birds have a cinnamon head and neck, and bold, patterns on their black-and-white wings and upperparts. The American Avocet forms large flocks during migration and in winter.

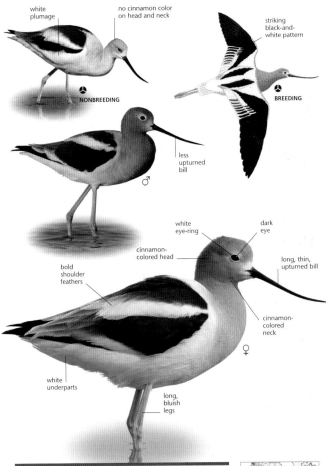

white plumage

no cinnamon color on head and neck

striking black-and-white pattern

NONBREEDING

BREEDING

less upturned bill

♂

white eye-ring

dark eye

cinnamon-colored head

long, thin, upturned bill

bold shoulder feathers

cinnamon-colored neck

white underparts

long, bluish legs

♀

VOICE *Variable melodic* kleet, *loud and repetitive, when alarmed and while foraging.*
NESTING *Simple, shallow scrape; 4 eggs; 1 brood; May–Jun.*
FEEDING *Uses specialized bill to probe or jab at a variety of aquatic invertebrates, small fish, and seeds; walks in belly-deep water to chase its prey.*
HABITAT *Temporary wetlands in dry to arid regions while breeding; shallow water habitats in winter.*
LENGTH *17–18½in (43–47cm)*
WINGSPAN *29–32in (74–81cm)*

Pacific Golden Plover

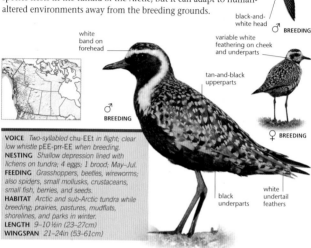

Pluvialis fulva

Although the Pacific Golden Plover frequents grassy habitats, it is also regularly encountered in coastal habitats as it migrates over the ocean to wintering grounds on remote South Pacific Islands. The species nests in the tundra of the Arctic, but it can adapt to human-altered environments away from the breeding grounds.

long wings

black-and-white head

♂ **BREEDING**

white band on forehead

variable white feathering on cheek and underparts

tan-and-black upperparts

♂ **BREEDING**

♀ **BREEDING**

VOICE *Two-syllabled chu-EEt in flight; clear low whistle pEE-prr-EE when breeding.*
NESTING *Shallow depression lined with lichens on tundra; 4 eggs; 1 brood; May–Jul.*
FEEDING *Grasshoppers, beetles, wireworms; also spiders, small mollusks, crustaceans, small fish, berries, and seeds.*
HABITAT *Arctic and sub-Arctic tundra while breeding; prairies, pastures, mudflats, shorelines, and parks in winter.*
LENGTH 9–10½in (23–27cm)
WINGSPAN 21–24in (53–61cm)

black underparts

white undertail feathers

American Golden Plover

Pluvialis dominica

The American Golden Plover is seen in North America only during its lengthy migrations to and from its high Arctic breeding grounds and wintering locations in southern South America. Its annual migration route includes a feeding stop at Labrador, followed by a 1,550–1,860 miles (2,500–3,000km) flight over the Atlantic Ocean to South America.

black-and-white face

dark tail

BREEDING

dark cap

brownish upperparts

small, thin bill

uniformly dusky underparts

NONBREEDING

white stripe from forehead to nape

tan-and-black spangled upperparts

BREEDING

black underparts

black legs

VOICE *Whistled two-note queE-dle or klee-u in flight; males: strong, melodious whistled kid-eek, or kid-EEp in flight.*
NESTING *Shallow depression lined with lichens in dry tundra; 4 eggs; 1 brood; May–Jul.*
FEEDING *Insects, mollusks, crustaceans, and worms; also berries and seeds.*
HABITAT *Arctic tundra while breeding; prairies, farmlands, golf course, pastures, mudflats, and shorelines while migrating.*
LENGTH 9½–11in (24–28cm)
WINGSPAN 23–28in (59–72cm)

Black-bellied Plover

Pluvialis squatarola

The Black-bellied Plover is the largest and most common of the
three North American *Pluvialis* plovers. Its preference for open
feeding habitats, its bulky structure, and very upright stance
make it a fairly conspicuous species. The Black-bellied Plover's
black underwing patches, visible in flight, are present in both
its breeding and nonbreeding plumages and distinguish it
from the other *Pluvialis* plovers, along with its larger size.

white
rump

♂ BREEDING

white
wing
stripe

darker
crown

diffused
streaks to
upper breast

whitish
underparts

NONBREEDING

duller plumage
than male

♀ MOLTING TO
BREEDING PLUMAGE

whitish
crown

checkered,
black-and-
white upperparts

black
cheeks

black
belly

♂ BREEDING

VOICE *Clear, plaintive, whistled whEE-er-eee, with middle
note lower; breeding males: softer, with accent on second syllable.*
NESTING *Shallow depression lined with mosses and lichens in
moist to dry lowland tundra; 1–5 eggs; 1 brood; May–Jul.*
FEEDING *Forages along coasts in typical run, pause, and pluck
style; insects, worms, bivalves, and crustaceans.*
HABITAT *High Arctic habitats while breeding; coastal areas
in winter.*
LENGTH *10½–12in (27–30cm)*
WINGSPAN *29–32in (73–81cm)*

Semipalmated Plover (S)

Charadrius semipalmatus

The Semipalmated Plover is a small bird with a tapered shape. It is common in a wide variety of habitats during migration and in winter, when these birds gather in loose flocks. Walking down a sandy beach between fall and spring might awaken up to 100 Semipalmated Plovers, sleeping in slight depressions in the sand; flocks may number up to 1,000 birds.

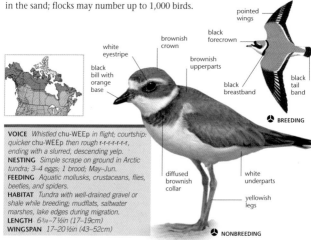

pointed wings

black forecrown

brownish crown

white eyestripe

black bill with orange base

brownish upperparts

black breastband

black tail band

🏵 BREEDING

diffused brownish collar

white underparts

yellowish legs

🏵 NONBREEDING

VOICE *Whistled chu-WEEp in flight; courtship: quicker chu-WEEp then rough r-r-r-r-r-r, ending with a slurred, descending yelp.*
NESTING *Simple scrape on ground in Arctic tundra; 3–4 eggs; 1 brood; May–Jun.*
FEEDING *Aquatic mollusks, crustaceans, flies, beetles, and spiders.*
HABITAT *Tundra with well-drained gravel or shale while breeding; mudflats, saltwater marshes, lake edges during migration.*
LENGTH *6¾–7½in (17–19cm)*
WINGSPAN *17–20½in (43–52cm)*

Killdeer (S)

Charadrius vociferus

This loud and vocal shorebird is the most widespread plover in North America, nesting in all southern Canadian provinces and across the US. Its piercing call carries for long distances, sometimes causing other birds to fly away in fear of imminent danger. These birds often nest near human habitation, allowing a close observation of their vigilant parental nature with young chicks.

white wing bar

long wings

reddish orange tail and rump

brownish crown

red eye-ring

second neck band crosses upper breast

brownish upperparts

small, thin, black bill

rufous wash to back and wings

long tail

white underparts

pinkish legs, sometimes with yellowish tinge

♂

black collar encircling neck

VOICE *Rising, drawn out deeee in flight; loud dee-ee in alarm, given repetitively; series of dee notes, followed by rising trill in agitation.*
NESTING *Scrape on ground; 4 eggs; 1 brood (north); Mar–Jul.*
FEEDING *Run, pause, and pluck foraging; worms, snails, grasshoppers, and beetles; also small vertebrates and seeds.*
HABITAT *Shorelines, mudflats, lakes, sparsely grassy fields, golf courses, and roadsides.*
LENGTH *9–10in (23–26cm)*
WINGSPAN *23–25in (58–63cm)*

Piping Plover

Charadrius melodus

With its pale gray back, the Piping Plover is well camouflaged along beaches and in dunes where it nests. Two subspecies of the Piping Plover are recognized; one nests on the Atlantic Coast, and the other inland. It is at risk due to eroding coastlines, human disturbance, and predation by foxes, raccoons, and cats.

stubby bill

prominent white wing stripe

dusky tail band

♂ **BREEDING**

black forecrown

dark breastband

pale gray upperparts

black-tipped, orange bill

♂ **BREEDING**

orange legs

thin, white collar throughout the year

mostly black bill, with slight orange base

indistinct, partial breastband

NONBREEDING

VOICE *Clear, whistled peep call in flight; quiet peep-lo during courtship and contact; high-pitched pipe-pipe-pipe song.*
NESTING *Shallow scrape in sand, gravel, dunes, or salt flats; 4 eggs; 1 brood; Apr–May.*
FEEDING *Run, pause, and pluck foraging; marine worms, insects, and mollusks.*
HABITAT *Beaches, saline sandflats, and mudflats; inland species nest in sand or gravel beaches by large lakes and rivers.*
LENGTH *6½–7 in (17–18cm)*
WINGSPAN *18–18½ in (45–47cm)*

American Woodcock

Scolopax minor

This forest-dwelling member of the sandpiper family bears little resemblance in behavior to its water-favoring relatives, but slightly resembles Wilson's Snipe and the dowitchers. Although widespread, the American Woodcock is largely nocturnal and seldom seen, except during its impressive twilight courtship displays. It feeds in mature fields or woodlands. Its noisy, repetitive display flights are a sign of spring in northern breeding areas.

short, rusty tail

two pale bands across back

large, black eye

long bill, wide at base with slightly drooping tip

black, gray, and buff upperparts

rich orange-buff underparts

round, plump body

short, rusty tail

pinkish legs and feet

VOICE *Males: Low, nasal peen during dawn and dusk display; wings make chirping and twittering sounds in display flight.*
NESTING *Shallow depression in leaf and twig litter in young, mixed growth woodlands; 4 eggs; 1 brood; Apr (North).*
FEEDING *Probes for earthworms, but also eats insects, snails, and some plants.*
HABITAT *Damp, second growth forest, overgrown fields, and bogs while breeding.*
LENGTH *10–12 in (25–31cm)*
WINGSPAN *16–20 in (41–51cm)*

Wilson's Snipe (S)

Gallinago delicata

Also known as the Common Snipe, Wilson's Snipe is a well-camouflaged member of the sandpiper family. On its breeding grounds, it produces eerie sounds during its aerial, mainly nocturnal, display flights. The birds fly up silently from the ground and from about 330 feet (100m) up, they descend quickly with tail feathers spread, producing a loud and vibrating sound through modified feathers.

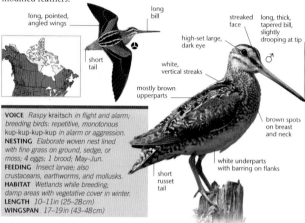

long, pointed, angled wings

long bill

short tail

streaked face

long, thick, tapered bill, slightly drooping at tip

high-set large, dark eye

♂

white, vertical streaks

mostly brown upperparts

brown spots on breast and neck

white underparts with barring on flanks

short russet tail

VOICE *Raspy* kraitsch *in flight and alarm; breeding birds: repetitive, monotonous* kup-kup-kup-kup *in alarm or aggression.*
NESTING *Elaborate woven nest lined with fine grass on ground, sedge, or moss; 4 eggs; 1 brood; May–Jun.*
FEEDING *Insect larvae; also crustaceans, earthworms, and mollusks.*
HABITAT *Wetlands while breeding; damp areas with vegetative cover in winter.*
LENGTH *10–11in (25–28cm)*
WINGSPAN *17–19in (43–48cm)*

Short-billed Dowitcher (S)

Limnodromus griseus

The Short-billed Dowitcher is a common visitor along the Atlantic, Gulf, and Pacific Coasts. There are three subspecies (*L. g. griseus*, *L. g. hendersoni*, and *L. g. caurinus*), which differ in plumage, size, and respective breeding areas. Birders should use extreme caution when differentiating them from Long-billed Dowitchers during migration and on wintering grounds.

slightly larger bill

L. g. hendersoni

long, pointed wings

white slash from rump to mid-back

orange wash to face, neck, breast, and underparts

BREEDING

long, stout bill

dark-centered upperpart feathers

variable spotting on upper breast

streaked flanks

greenish yellow legs

L. g. griseus

VOICE *Low* tu-tu-tu *or* tu-tu, tu-tu, toodle-ee, tu-tu, *ending with low* anh-anh-anh *in flight.*
NESTING *Simple depression, typically in sedge hummock; 4 eggs; 1 brood; May–Jun.*
FEEDING *Probes in "sewing machine" style in belly-deep water for mollusks, crustaceans, and insects.*
HABITAT *Sedge meadows or bogs with spruce and tamaracks while breeding; coastal mudflats and salt marshes in winter.*
LENGTH *9–10in (23–25cm)*
WINGSPAN *18–20in (46–51cm)*

Long-billed Dowitcher Ⓢ

Limnodromus scolopaceus

The Long-billed Dowitcher is usually slightly larger, longer-legged, and heavier in the chest and neck than the Short-billed Dowitcher. The breeding ranges of the two species are separate, but their migration and en route stop-over areas overlap. The Long-billed Dowitcher is usually found in freshwater wetlands, and in the fall most of its population occurs west of the Mississippi River.

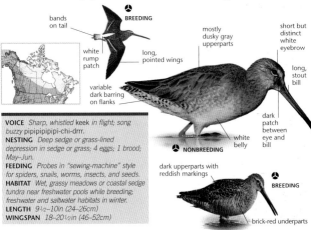

bands on tail

Ⓑ BREEDING

white rump patch

long, pointed wings

variable dark barring on flanks

mostly dusky gray upperparts

short but distinct white eyebrow

long, stout bill

dark patch between eye and bill

white belly

Ⓝ NONBREEDING

dark upperparts with reddish markings

Ⓑ BREEDING

brick-red underparts

VOICE *Sharp, whistled* keek *in flight; song buzzy* pipipipipipi-chi-drrr.
NESTING *Deep sedge or grass-lined depression in sedge or grass; 4 eggs; 1 brood; May–Jun.*
FEEDING *Probes in "sewing-machine" style for spiders, snails, worms, insects, and seeds.*
HABITAT *Wet, grassy meadows or coastal sedge tundra near freshwater pools while breeding; freshwater and saltwater habitats in winter.*
LENGTH 9½–10in (24–26cm)
WINGSPAN 18–20½in (46–52cm)

Hudsonian Godwit Ⓢ

Limosa haemastica

This sandpiper undertakes a remarkable annual migration from its tundra breeding grounds in Alaska and Canada all the way to extreme southern South America, a distance probably close to 10,000 miles (16,000km) in one direction, with very few stopovers. During migration, North American stops are few and occur only in the spring, along a central route mid-continent.

white wing stripe

white rump

Ⓝ NONBREEDING

brownish streaked head and neck

white-feathered chestnut breast

♀ BREEDING

black-and-white upperparts

long, orange-based bill

unpatterned brownish wing feathers

black tail

rich chestnut underparts with black barring

♂ BREEDING

VOICE *Emphatic* peed-wid *in flight; high* peet *or* kwee; *display song* to-wida to-wida to-wida, *or* to-wit, to-wit, to-wit.
NESTING *Saucer-shaped depression on dry hummock; 4 eggs; 1 brood; May–Jul.*
FEEDING *Probes for insects, grubs, worms, crustaceans, mollusks; plant tubers in fall.*
HABITAT *Sedge meadows and bogs in tundra while breeding; Atlantic Coast near freshwater in fall; flooded fields, reservoirs in spring.*
LENGTH 14–16in (35–41cm)
WINGSPAN 27–31in (68–78cm)

Marbled Godwit (S)

Limosa fedoa

The largest godwit in North America, this shorebird is a familiar sight at its coastal wintering areas. Its distinctive brown-and-cinnamon plumage and the fact that it chooses open habitats, such as mudflats and floodplains, to feed and roost, make the Marbled Godwit a conspicuous species.

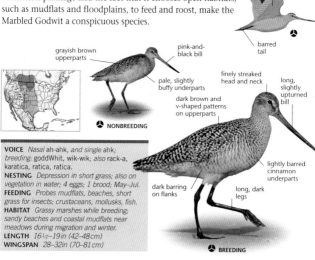

cinnamon underwing

barred tail

grayish brown upperparts

pink-and-black bill

pale, slightly buffy underparts

NONBREEDING

finely streaked head and neck

long, slightly upturned bill

dark brown and v-shaped patterns on upperparts

lightly barred cinnamon underparts

dark barring on flanks

long, dark legs

BREEDING

VOICE *Nasal ah-ahk, and single ahk; breeding: goddWhit, wik-wik; also rack-a, karatica, ratica, ratica.*
NESTING *Depression in short grass; also on vegetation in water; 4 eggs; 1 brood; May–Jul.*
FEEDING *Probes mudflats, beaches, short grass for insects; crustaceans, mollusks, fish.*
HABITAT *Grassy marshes while breeding; sandy beaches and coastal mudflats near meadows during migration and winter.*
LENGTH *16½–19in (42–48cm)*
WINGSPAN *28–32in (70–81cm)*

Whimbrel (S)

Numenius phaeopus

This large, conspicuous shorebird is the most widespread of the curlew species, with four subspecies across North America and Eurasia. Its bold head stripes and clearly streaked face, neck, and breast make the species distinctive. The Whimbrel's fairly long, decurved bill allows it to probe into fiddler crab burrows, a favorite food item.

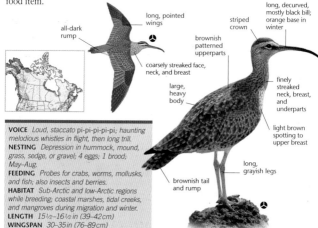

long, pointed wings

all-dark rump

coarsely streaked face, neck, and breast

striped crown

long, decurved, mostly black bill; orange base in winter

brownish patterned upperparts

large, heavy body

finely streaked neck, breast, and underparts

light brown spotting to upper breast

brownish tail and rump

long, grayish legs

VOICE *Loud, staccato pi-pi-pi-pi-pi; haunting melodious whistles in flight, then long trill.*
NESTING *Depression in hummock, mound, grass, sedge, or gravel; 4 eggs; 1 brood; May–Aug.*
FEEDING *Probes for crabs, worms, mollusks, and fish; also insects and berries.*
HABITAT *Sub-Arctic and low-Arctic regions while breeding; coastal marshes, tidal creeks, and mangroves during migration and winter.*
LENGTH *15½–16½in (39–42cm)*
WINGSPAN *30–35in (76–89cm)*

Long-billed Curlew

D

Numenius americanus

The Long-billed Curlew has the southernmost breeding range and northern most wintering range of the four North American curlews. It is also one of nine species of birds that are endemic to the grasslands of the Great Plains. The downward curvature of its extremely long bill is adapted to probe for food in soft mud and sand.

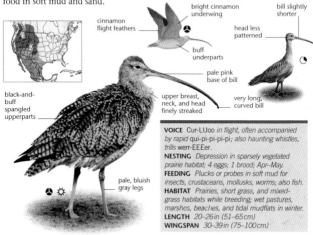

bright cinnamon underwing

bill slightly shorter

cinnamon flight feathers

head less patterned

buff underparts

pale pink base of bill

black-and-buff spangled upperparts

upper breast, neck, and head finely streaked

very long, curved bill

pale, bluish gray legs

VOICE *Cur-LUoo in flight, often accompanied by rapid qui-pi-pi-pi-pi; also haunting whistles, trills werr-EEEer.*
NESTING *Depression in sparsely vegetated prairie habitat; 4 eggs; 1 brood; Apr–May.*
FEEDING *Plucks or probes in soft mud for insects, crustaceans, mollusks, worms; also fish.*
HABITAT *Prairies, short grass, and mixed-grass habitats while breeding; wet pastures, marshes, beaches, and tidal mudflats in winter.*
LENGTH *20–26in (51–65cm)*
WINGSPAN *30–39in (75–100cm)*

Upland Sandpiper

S

Bartramia longicauda

Unlike other sandpipers, the graceful Upland Sandpiper spends most of its life away from water in grassland habitats where its coloration helps it camouflage itself, especially while nesting in the grass. It is well known for landing on wooden fence posts and raising its wings while giving its tremulous, whistling call.

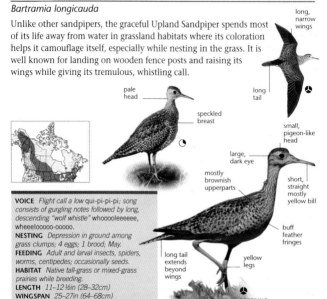

long, narrow wings

long tail

pale head

speckled breast

small, pigeon-like head

large, dark eye

short, straight mostly yellow bill

mostly brownish upperparts

buff feather fringes

long tail extends beyond wings

yellow legs

BREEDING

VOICE *Flight call a low qui-pi-pi-pi; song consists of gurgling notes followed by long, descending "wolf whistle" whooooleeeeee, wheeelooooo-ooooo.*
NESTING *Depression in ground among grass clumps; 4 eggs; 1 brood; May.*
FEEDING *Adult and larval insects, spiders, worms, centipedes; occasionally seeds.*
HABITAT *Native tall-grass or mixed-grass prairies while breeding.*
LENGTH *11–12½in (28–32cm)*
WINGSPAN *25–27in (64–68cm)*

Greater Yellowlegs

(S)

Tringa melanoleuca

This fairly large shorebird often runs frantically in many directions
while pursuing small prey. It is one of the first northbound spring
shorebird migrants, and one of the first to return south in late June
or early July. Its plumage is more streaked during the breeding
season and its bill is slightly upturned.

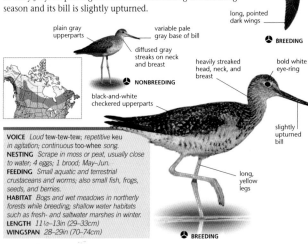

long, pointed
dark wings

BREEDING

plain gray
upperparts

variable pale
gray base of bill

diffused gray
streaks on neck
and breast

NONBREEDING

black-and-white
checkered upperparts

heavily streaked
head, neck, and
breast

bold white
eye-ring

slightly
upturned
bill

long,
yellow
legs

BREEDING

VOICE *Loud tew-tew-tew; repetitive keu
in agitation; continuous too-whee song.*
NESTING *Scrape in moss or peat, usually close
to water; 4 eggs; 1 brood; May–Jun.*
FEEDING *Small aquatic and terrestrial
crustaceans and worms; also small fish, frogs,
seeds, and berries.*
HABITAT *Bogs and wet meadows in northerly
forests while breeding; shallow water habitats
such as fresh- and saltwater marshes in winter.*
LENGTH *11½–13in (29–33cm)*
WINGSPAN *28–29in (70–74cm)*

Lesser Yellowlegs

(S)

Tringa flavipes

The Lesser Yellowlegs has a smaller head, thinner bill, and
smoother body shape than the Greater Yellowlegs. It prefers
smaller, freshwater, or brackish pools to open saltwater
habitats, and it walks quickly and methodically while feeding.
Although this species is a solitary feeder, it is often seen in
small to large loose flocks in migration and winter.

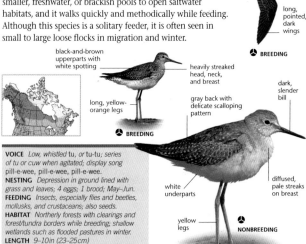

small
head

long,
pointed,
dark
wings

BREEDING

black-and-brown
upperparts with
white spotting

heavily streaked
head, neck,
and breast

dark,
slender
bill

long, yellow-
orange legs

gray back with
delicate scalloping
pattern

BREEDING

white
underparts

diffused,
pale streaks
on breast

yellow
legs

NONBREEDING

VOICE *Low, whistled tu, or tu-tu; series
of tu or cuw when agitated; display song
pill-e-wee, pill-e-wee, pill-e-wee.*
NESTING *Depression in ground lined with
grass and leaves; 4 eggs; 1 brood; May–Jun.*
FEEDING *Insects, especially flies and beetles,
mollusks, and crustaceans; also seeds.*
HABITAT *Northerly forests with clearings and
forest/tundra borders while breeding; shallow
wetlands such as flooded pastures in winter.*
LENGTH *9–10in (23–25cm)*
WINGSPAN *23–25in (58–64cm)*

Spotted Sandpiper

Ⓢ

Actitis macularius

This small, short-legged sandpiper is the most widespread shorebird in North America. It is characterized by its quick walking pace, its habit of constantly teetering and bobbing its tail, and its unique style of flying low over water. Spotted Sandpipers have an unusual mating behavior, in which the females take on an aggressive role, defending territories and serially mating with three or more males per season.

white wing stripe

darker flight feathers

Ⓑ **BREEDING**

plain brownish-gray upperparts

straight, dark bill

white wedge on breast

Ⓑ **NONBREEDING**

brownish gray upperparts

bold, white eye-ring

thin, white eyestripe

buff barring on wings and back

straight, orange bill with dark tip

brownish gray upperparts

dark barring on back

Ⓑ **BREEDING**

white underparts with bold, dark spots

orange-yellow legs

VOICE *Clear, ringing* tee-tee-tee-tee; *monotonous* cree-cree-cree *in flight.*
NESTING *Nest cup shaded by or scrape built under herbaceous vegetation; 3 eggs; 1–3 broods; May–Jun.*
FEEDING *Many items, including adult and larval insects, mollusks, small crabs, and worms.*
HABITAT *Variety of grassy, brushy, forested habitats near water while breeding; freshwater habitats (lakeshores, rivers, beaches) in winter.*
LENGTH *7¼–8in (18.5–20cm)*
WINGSPAN *15–16in (38–41cm)*

Solitary Sandpiper

Tringa solitaria

This aptly named Sandpiper seldom associates with other shorebirds as it moves nervously along margins of wetlands. When feeding, it constantly bobs its head like the Spotted Sandpiper. When disturbed, the Solitary Sandpiper often flies directly upward, and when landing, it keeps its wings upright briefly, flashing the white underneath, before carefully folding them to its body.

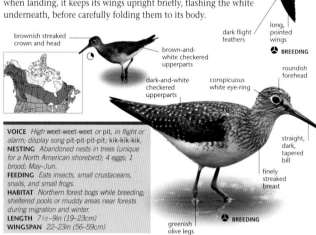

dark flight feathers

long, pointed wings

BREEDING

brownish streaked crown and head

brown-and-white checkered upperparts

dark-and-white checkered upperparts

roundish forehead

conspicuous white eye-ring

straight, dark, tapered bill

finely streaked breast

greenish olive legs

BREEDING

VOICE *High weet-weet-weet or pit, in flight or alarm; display song pit-pit-pit-pit; kik-kik-kik.*
NESTING *Abandoned nests in trees (unique for a North American shorebird); 4 eggs; 1 brood; May–Jun.*
FEEDING *Eats insects, small crustaceans, snails, and small frogs.*
HABITAT *Northern forest bogs while breeding; sheltered pools or muddy areas near forests during migration and winter.*
LENGTH *7½–9in (19–23cm)*
WINGSPAN *22–23in (56–59cm)*

Wandering Tattler

Tringa incana

The Wandering Tattler is named for its long annual migration and its loud calls, heard in its breeding sites in Alaska and western Canada. Seen singly or occasionally in small groups on the rocky Pacific Coast shoreline from late summer to spring, this mostly solitary species is often overlooked.

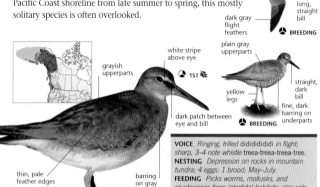

long, pointed wings

fairly long, straight bill

dark gray flight feathers

BREEDING

grayish upperparts

white stripe above eye

plain gray upperparts

1ST

yellow legs

dark patch between eye and bill

straight, dark bill

fine, dark barring on underparts

BREEDING

thin, pale feather edges

barring on gray breast

dull yellow-green legs

VOICE *Ringing, trilled dididididi in flight; sharp, 3–4 note whistle treea-treea-treea-tree.*
NESTING *Depression on rocks in mountain tundra; 4 eggs; 1 brood; May–July.*
FEEDING *Picks worms, mollusks, and crustaceans from intertidal habitats; also eats insects, sand fleas, and fish.*
HABITAT *Shrubby mountainous Arctic tundra close to water while breeding; rocky coastlines during migration and winter.*
LENGTH *10½–12in (27–30cm)*
WINGSPAN *20–22in (51–56cm)*

Willet

Tringa semipalmata

The two distinct subspecies of the Willet, Eastern *(T. s. semipalmatus)* and Western *(T. s. inornatus)*, differ in breeding habit, plumage coloration, vocalizations, and migratory habits. The Eastern Willet leaves North America from September to March, whereas the Western Willet winters along southern North American shorelines south to South America.

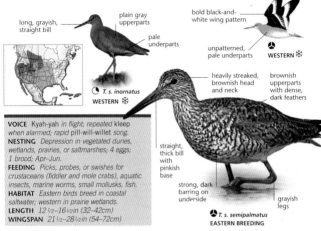

long, grayish, straight bill

plain gray upperparts

pale underparts

T. s. inornatus
WESTERN ❄

bold black-and-white wing pattern

unpatterned, pale underparts
WESTERN ❄

heavily streaked, brownish head and neck

brownish upperparts with dense, dark feathers

straight, thick bill with pinkish base

strong, dark barring on underside

grayish legs

T. s. semipalmatus
EASTERN BREEDING

VOICE Kyah-yah *in flight; repeated* kleep *when alarmed; rapid* pill-will-willet *song.*
NESTING *Depression in vegetated dunes, wetlands, prairies, or saltmarshes; 4 eggs; 1 brood; Apr–Jun.*
FEEDING *Picks, probes, or swishes for crustaceans (fiddler and mole crabs), aquatic insects, marine worms, small mollusks, fish.*
HABITAT *Eastern birds breed in coastal saltwater; western in prairie wetlands.*
LENGTH 12½–16½in (32–42cm)
WINGSPAN 21½–28½in (54–72cm)

Ruddy Turnstone

Arenaria interpres

A common visitor along the shorelines of North and South America, the Ruddy Turnstone is aggressive on its high-Arctic breeding grounds, driving off predators as large as the Glaucous Gull and Parasitic Jaeger. It was named after its reddish back color and its habit of overturning items like mollusk shells and pebbles, looking for small crustaceans and other prey.

brownish upperparts

variably streaked, whitish face

brownish head markings

NONBREEDING

dark flight feathers

bold red patches on back and wings
BREEDING

short, dark, chisel-like bill

black breast

black-and-white head and breast pattern

bright white underparts, at all ages

short, orange legs

BREEDING

VOICE TIT-wooo TIT-woooRITititititititit *on breeding ground; low, rapid* kut-a-kut *in flight.*
NESTING *Scrape lined with lichens and grasses in dry, open areas; 4 eggs; 1 brood; Jun.*
FEEDING *Forages along shoreline for crustaceans, insects, including beetles, spiders; also plants.*
HABITAT *Open, barren, grassy habitats near water and rocky coasts while breeding; beaches and rocky shorelines in winter.*
LENGTH 8–10½in (20–27cm)
WINGSPAN 20–22½in (51–57cm)

Black Turnstone

Arenaria melanocephala

The Black Turnstone blends in well on rocky shorelines, becoming almost invisible when it forages or roosts on dark, rocky surfaces. It uses its chisel-like bill to flip stones and beach litter in search of food, and to pry loose or crack tougher prey, particularly mussels and barnacles. This species is an aggressive defender of the nesting community, even physically attacking predators such as jaegers.

white patch on back

black tail band

stocky, pointed wings

NONBREEDING

blackish back

white patch

black head and breast with white flecking

darker legs

BREEDING

brownish upperparts, with scattered black feathers

short, blackish, chisel-like bill

dark chocolate-brown head and breast

white belly

pale edges to some feathers

NONBREEDING

yellowish legs

VOICE *Breerp* in flight, often continued as rapid chattering; trills, purrs, and a *tu-whit* call.
NESTING Hollow depression in tundra; 4 eggs; 1 brood; May–Jun.
FEEDING Invertebrates (mussels, limpets, snails, crabs); seeds, small bird eggs, carrion.
HABITAT Tundra, and inland along rivers and lakes while breeding; tidal zone of rocky shorelines, beaches, and rocky jetties during migration and winter.
LENGTH 8½–10½ in (22–27cm)
WINGSPAN 20–22½ in (51–57cm)

Surfbird

Aphriza virgata

The chunky, stubby-billed Surfbird breeds in the high mountain tundra of Alaska and the Yukon and then migrates to the rocky Pacific coasts of both North and South America. Some individuals migrate as far as southern Chile, a round trip of about 19,000 miles (30,500km) each year. The extent of the rust color on the upperparts of breeding Surfbirds is variable.

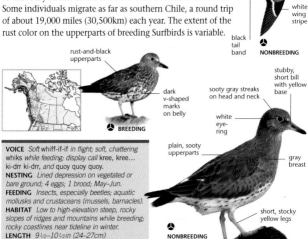

white wing stripe

black tail band

NONBREEDING

rust-and-black upperparts

dark v-shaped marks on belly

BREEDING

stubby, short bill with yellow base

sooty gray streaks on head and neck

white eye-ring

plain, sooty upperparts

gray breast

short, stocky yellow legs

NONBREEDING

VOICE *Soft whiff-if-if* in flight; soft, chattering *whiks* while feeding; display call *kree, kree… ki-drr ki-drr*, and *quoy quoy quoy*.
NESTING Lined depression on vegetated or bare ground; 4 eggs; 1 brood; May–Jun.
FEEDING Insects, especially beetles; aquatic mollusks and crustaceans (mussels, barnacles).
HABITAT Low to high-elevation steep, rocky slopes of ridges and mountains while breeding; rocky coastlines near tideline in winter.
LENGTH 9½–10½in (24–27cm)
WINGSPAN 25–27in (63–68cm)

Sanderling

S

Calidris alba

The Sanderling is probably the best-known shorebird in the world. It breeds in remote, high-Arctic habitats, from Greenland to Siberia, but occupies just about every temperate and tropical shoreline in the Americas when not breeding. Its wintering range spans both American coasts, from Canada to Argentina. Feeding flocks, constantly on the move at the water's edge, are a common sight in winter on sandy beaches. In many places, the bird is declining rapidly, with pollution of the sea and shore and habitat disturbances the main causes.

mostly grayish upperparts

strong white wing stripe

NONBREEDING

white face and neck

pearl-gray upperparts

clean white underparts

NONBREEDING

black-centered back feathers with buff edges

rust and black streaked crown

black, rust, and white upperparts

dark, stocky bill

rust wash on breast with black markings

short black legs

BREEDING

VOICE *Squeaky* pweet *in flight,* sew-sew-sew *threat call; display song harsh, buzzy notes and chattering* cher-cher-cher.
NESTING *Small, shallow depression on dry, stony ground; 4 eggs; 1–3 broods; Jun–Jul.*
FEEDING *Probes along the surf-line in sand for insects, small crustaceans, small mollusks, and worms.*
HABITAT *Barren high-Arctic coastal tundra while breeding; coastlines and sandy beaches during migration and winter.*
LENGTH *7½–8in (19–20cm)*
WINGSPAN *16–18in (41–46cm)*

Red Knot

Calidris canutus

The Red Knot is the largest North American shorebird in the genus Calidris. There are two North American subspecies—*C. c. rufa* and *C. c. roselaari. C. c. rufa* flies about 9,300 miles (15,000km) between its high-Arctic breeding grounds and wintering area in South America. Recent declines have occurred in this population, attributed to over-harvesting of horseshoe crab eggs—its critical food source.

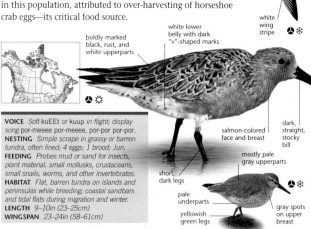

white eyebrow

white wing stripe

boldly marked black, rust, and white upperparts

white lower belly with dark "v"-shaped marks

salmon-colored face and breast

dark, straight, stocky bill

mostly pale gray upperparts

short, dark legs

pale underparts

yellowish green legs

gray spots on upper breast

VOICE *Soft kuEEt or kuup in flight; display song por-meeee por-meeee, por-por por-por.*
NESTING *Simple scrape in grassy or barren tundra, often lined; 4 eggs; 1 brood; Jun.*
FEEDING *Probes mud or sand for insects, plant material, small mollusks, crustaceans, small snails, worms, and other invertebrates.*
HABITAT *Flat, barren tundra on islands and peninsulas while breeding; coastal sandbars and tidal flats during migration and winter.*
LENGTH 9–10in (23–25cm)
WINGSPAN 23–24in (58–61cm)

Semipalmated Sandpiper

Calidris pusilla

This abundant sandpiper breeds in Canada's Arctic tundra. Flocks of up to 300,000 birds gather on migration staging areas. It can be hard to identify, due to plumage variation between juveniles and breeding adults, and a bill that varies in size and shape from west to east. Semipalmated sandpipers from northeasterly breeding grounds may fly nonstop to their South American wintering grounds in the fall.

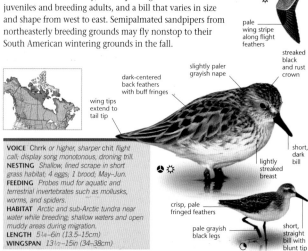

white eyebrow

pale wing stripe along flight feathers

streaked black and rust crown

slightly paler grayish nape

dark-centered back feathers with buff fringes

wing tips extend to tail tip

short, dark bill

lightly streaked breast

crisp, pale fringed feathers

pale grayish black legs

short, straight bill with blunt tip

VOICE *Chrrk or higher, sharper chit flight call; display song monotonous, droning trill.*
NESTING *Shallow, lined scrape in short grass habitat; 4 eggs; 1 brood; May–Jun.*
FEEDING *Probes mud for aquatic and terrestrial invertebrates such as mollusks, worms, and spiders.*
HABITAT *Arctic and sub-Arctic tundra near water while breeding; shallow waters and open muddy areas during migration.*
LENGTH 5¼–6in (13.5–15cm)
WINGSPAN 13½–15in (34–38cm)

Western Sandpiper

Calidris mauri

During its spring migration, around two million Western Sandpipers stop at the mudflats of Roberts Bank in British Columbia, to refuel for the last hop northward to their Western Alaska breeding grounds. Many of these migrate over relatively short distances to winter along US coastlines, so the timing of their molt in fall is earlier than that of the similar Semipalmated Sandpiper, which migrates later in winter.

white tail

dusky tail band

long, narrow, pointed wing

narrow, white wing stripe

mostly uniform brown or grayish upperparts

grayish, streaked crown, nape, and face

dark patch between eyes and bill

NONBREEDING

partial grayish, streaked collar

white belly

bright, rusty cap and cheek patch

BREEDING

medium-length black legs

grayish, streaked nape and neck

VOICE Loud *chir-eep* in flight; *sirp* or *chir-ir-ip* when flushed; *tweer, tweer, tweer,* followed by descending trill.
NESTING *Depression on drained Arctic and sub-Arctic tundra; 4 eggs; 1 brood; May–Jun.*
FEEDING *In "sewing machine" style, probes mud for insect larvae, crustaceans, and worms.*
HABITAT *Wet sedge, grassy habitats with well-drained areas while breeding; shallow waters with open muddy or sandy areas in winter.*
LENGTH 5½–6½in (14–16cm)
WINGSPAN 14–15in (35–38cm)

Least Sandpiper

Calidris minutilla

With its muted, brown or brownish gray plumage, the tiny Least Sandpiper virtually disappears in the landscape when feeding crouched down on wet margins of water bodies. These birds are often found in small to medium flocks at the edge of other shorebird flocks. They are often nervous when foraging, and frequently burst into flight, only to alight a short way off.

faint tail band

uniform brownish gray upperparts

pale, whitish eyebrow

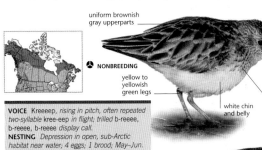

NONBREEDING

yellow to yellowish green legs

white chin and belly

streaked, brownish breast and head

small, rounded head

VOICE *Kreeeep, rising in pitch, often repeated two-syllable kree-eep in flight; trilled b-reeee, b-reeee, b-reeee display call.*
NESTING *Depression in open, sub-Arctic habitat near water; 4 eggs; 1 brood; May–Jun.*
FEEDING *Small terrestrial and aquatic prey, especially sand fleas, mollusks, and flies.*
HABITAT *Wet, low-Arctic areas while breeding; muddy areas on lakeshores, riverbanks, fields, and tidal flats during migration and winter.*
LENGTH 4¾in (12cm)
WINGSPAN 13–14in (33–35cm)

short tail and wings

short, yellowish legs

BREEDING

White-rumped Sandpiper (S)

Calidris fuscicollis

The White-rumped Sandpiper has one of the longest migrations of any bird in the Western Hemisphere. From its High Arctic breeding grounds, it migrates in several long jumps to extreme southern South America—about 9,000–12,000 miles (14,500–19,300km), twice a year. Almost the entire population migrates through the central US and Canada in spring, with several stopovers. Its insect-like call and white rump aid identification.

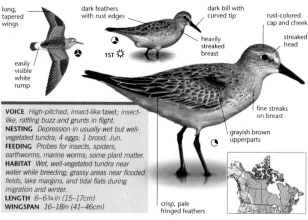

long, tapered wings

easily visible white rump

1ST

dark feathers with rust edges

dark bill with curved tip

heavily streaked breast

rust-colored cap and cheek

streaked head

fine streaks on breast

grayish brown upperparts

crisp, pale fringed feathers

VOICE High-pitched, insect-like tzeet; insect-like, rattling buzz and grunts in flight.
NESTING Depression in usually wet but well-vegetated tundra; 4 eggs; 1 brood; Jun.
FEEDING Probes for insects, spiders, earthworms, marine worms; some plant matter.
HABITAT Wet, well-vegetated tundra near water while breeding; grassy areas near flooded fields, lake margins, and tidal flats during migration and winter.
LENGTH 6–6¾in (15–17cm)
WINGSPAN 16–18in (41–46cm)

Baird's Sandpiper (S)

Calidris bairdii

Baird's Sandpiper is less well known than the other North American *Calidris* sandpipers. From its High Arctic, tundra habitat, Baird's Sandpiper migrates across western North America into South America, and all the way to Tierra del Fuego, a remarkable biannual journey of 6,000–9,000 miles (9,700–14,500km).

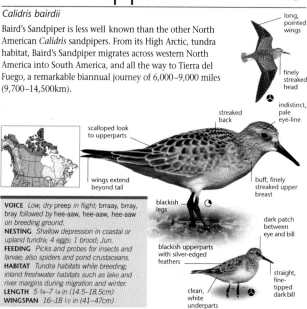

long, pointed wings

finely streaked head

indistinct, pale eye-line

streaked back

scalloped look to upperparts

wings extend beyond tail

blackish legs

buff, finely streaked upper breast

dark patch between eye and bill

blackish upperparts with silver-edged feathers

clean, white underparts

straight, fine-tipped dark bill

VOICE Low, dry preep in flight; brraay, brray, bray followed by hee-aaw, hee-aaw, hee-aaw on breeding ground.
NESTING Shallow depression in coastal or upland tundra; 4 eggs; 1 brood; Jun.
FEEDING Picks and probes for insects and larvae; also spiders and pond crustaceans.
HABITAT Tundra habitats while breeding; inland freshwater habitats such as lake and river margins during migration and winter.
LENGTH 5¾–7¼in (14.5–18.5cm)
WINGSPAN 16–18½in (41–47cm)

Pectoral Sandpiper

§

Calidris melanotos

From their breeding grounds in the High Arctic to their wintering grounds in South America, some Pectoral Sandpipers travel up to 30,000 miles (48,000km) each year. Males keep harems of females in guarded territories and mate with as many as they can attract, but take no part in nest duties and migrate earlier than females.

long, graceful, pointed wings

darker flight feathers

brownish upperparts, with buff fringes

streaked crown and face

curved bill with orange base

medium length, stocky bill

heavily streaked breast

rust crown and cheeks with black streaks

white belly

yellowish legs

rust-edged, dark centered feathers

VOICE Low, trilled chrrk in flight; deep, hollow whoop, whoop, whoop in display.
NESTING Shallow depression on ridges in moist to wet sedge tundra; 4 eggs; 1 brood; Jun.
FEEDING Probes or jabs mud for larvae, and forages for insects and spiders on tundra.
HABITAT Wet, grassy tundra near coasts while breeding; wet pastures, grassy lake margins, and saltmarshes during migration and winter.
LENGTH 7½–9in (19–23cm)
WINGSPAN 16½–19½in (42–49cm)

Purple Sandpiper

§

Calidris maritima

A medium-sized, stocky bird, the Purple Sandpiper shares the most northerly wintering distribution of all North American shorebirds with its close relative, the Rock Sandpiper. The dark plumage and low, squat body of the Purple Sandpiper often disguise its presence on dark tidal rocks, until a crashing wave causes a previously invisible flock to explode into flight.

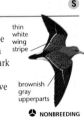

thin white wing stripe

brownish gray upperparts

NONBREEDING

grayish wash to head and neck

compact body shape overall

gray inner wing feathers

bill yellow at base, dark at drooping tip

white belly and flanks, with thin streaking

yellow legs and toes

NONBREEDING

long bill with drooping tip

heavily streaked head

short, thick neck

BREEDING

VOICE Low kweesh in flight; eh-eh-eh when disturbed; breeding kwi-ti-ti-ti-bli-bli-bli followed by dooree-dooree-dooree.
NESTING Simple lined scrape in Arctic tundra; 4 eggs; 1 brood; Jun.
FEEDING Invertebrates, including crustaceans, snails, insects, spiders, and worms.
HABITAT Barren Arctic and alpine tundra while breeding; rocky, wave-pounded shores during migration and winter.
LENGTH 8–8½in (20–21cm)
WINGSPAN 16½–18½in (42–47cm)

Rock Sandpiper

(S)

Calidris ptilocnemis

All three regularly occurring North American subspecies of this
bird breed in the Bering Sea region. The Rock Sandpiper is the
western, and closely related, counterpart of the Purple Sandpiper,
and the two species have the most northerly wintering range
of any shorebird in North America. Only one subspecies,
C. p. tschuktschorum, migrates to the Pacific coast of
North America.

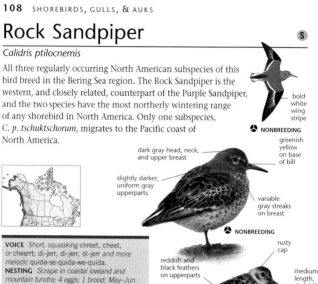

bold
white
wing
stripe

NONBREEDING

dark gray head, neck,
and upper breast

greenish
yellow
on base
of bill

slightly darker,
uniform gray
upperparts

variable
gray streaks
on breast

NONBREEDING

reddish and
black feathers
on upperparts

rusty
cap

medium
length,
dark bill

variable black
belly patch

C. p. ptilocnemis
PRIBILOF; BREEDING

VOICE *Short, squeaking* chreet, cheet,
or cheerrt; di-jerr, di-jerr, di-jerr *and more
melodic* quida-se-quida-we-quida.
NESTING *Scrape in coastal lowland and
mountain tundra; 4 eggs; 1 brood; May–Jun.*
FEEDING *Probes for clams and snails in
seaweed; land insects (beetles) while breeding.*
HABITAT *Arctic lowland coastal heath or
mountain tundra while breeding; rocky shores,
beaches, mud- and sandflats in winter.*
LENGTH *7¼–9½ in (18.5–24cm)*
WINGSPAN *13–18½ in (33–47cm)*

Dunlin

(S)

Calidris alpina

The Dunlin is one of the most abundant and widespread of North America's
shorebirds. Three subspecies breed in North America: *C. a. arcticola*, *C. a. pacifica*,
and *C. a. hudsonia*. The Dunlin is unmistakable in its striking, red-backed, black-
bellied breeding plumage. In winter it sports much more drab colors, but is
conspicuous by gathering in flocks of many thousands of birds on coastal mudflats.

thin white
wing bar

long, tapered,
black bill

dull gray-brown
head and back

white
sided
rump

dull, gray-
streaked
breast

NONBREEDING

rich chestnut-
and-black back

fine dark
streaks
on
whitish
breast

large, squarish,
black belly patch

BREEDING

VOICE *Accented trill,* drurr-drurr;
jeeezp *in flight;* wrraah-wrraah *song.*
NESTING *Cup lined with grasses and lichens
in moist tundra; 4 eggs; 1 brood; Jun–Jul.*
FEEDING *Marine, freshwater, terrestrial
invertebrates: clams, worms, insect larvae,
crustaceans; also plants and small fish.*
HABITAT *Wet tundra near ponds while
breeding; coastal areas with mudflats and
beaches during migration and winter.*
LENGTH *6½–8½ in (16–22cm)*
WINGSPAN *12½–17½ in (32–44cm)*

Stil Sandpiper

S

Calidris himantopus

The Stilt Sandpiper breeds in several small areas of northern tundra. It feeds by walking slowly through belly-deep water with its neck outstretched and bill pointed downward, either picking at the surface or submerging itself, keeping its tail raised up all the while. During migration it forms rapidly moving flocks that sometimes include other sandpiper species.

white rump

dusky tail band

long, pointed wing

NONBREEDING

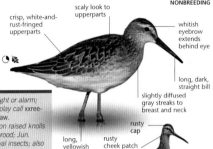

scaly look to upperparts

crisp, white-and-rust-fringed upperparts

whitish eyebrow extends behind eye

long, dark, straight bill

slightly diffused gray streaks to breast and neck

long, yellowish legs

rusty cheek patch

rusty cap

long wings and tail

chocolate-brown barring on white underparts

BREEDING

VOICE *Low, muffled chuf in flight or alarm; also krrit and sharp kew-it; display call xxree-xxree-xxree-xxree-ee-haw, ee-haw.*
NESTING *Shallow depression on raised knolls or ridges in tundra; 4 eggs; 1 brood; Jun.*
FEEDING *Mostly adult and larval insects; also some snails, mollusks, and seeds.*
HABITAT *Moist to wet tundra while breeding; freshwater habitats such as flooded fields and marsh pools during migration and winter.*
LENGTH *8–9in (20–23cm)*
WINGSPAN *17–18½in (43–47cm)*

Buff-breasted Sandpiper

D

Tryngites subruficollis

This sandpiper has a unique mating system among North American shorebirds. On the ground in the Arctic, each male flashes his white underwings to attract females. After mating, the female leaves to perform all nest duties alone, while the male continues to display and mate with other females. These birds migrate an astonishing 16,000 miles (26,000km) from their breeding grounds to winter in temperate South America.

pale central band

dark rump

NONBREEDING

streaked and spotted brown hind neck

buff-edged brown upperparts

buff head and face with spotted brown crown

short, dark bill

bright yellowish orange legs

BREEDING

scaly upperparts

more white-fringed upperpart feathers than adult

rich buff wash to breast

dull, yellow legs

VOICE *Soft, short gert in flight, or longer, rising grriit.*
NESTING *Simple depression on well-drained moss or grass hummock; 4 eggs; 1 brood; Jun.*
FEEDING *Forages on land for insects, insect larvae, and spiders; occasionally seeds.*
HABITAT *Moist to wet, grassy or sedge coastal tundra while breeding; short grass areas such as pastures and meadows during migration.*
LENGTH *7¼–8in (18.5–20cm)*
WINGSPAN *17–18½in (43–47cm)*

Wilson's Phalarope

S

Phalaropus tricolor

The largest of the three phalarope species, Wilson's Phalarope breeds in the shallow wetlands of western North America and winters mainly in Bolivia and Argentina. It feeds by spinning in shallow water to churn up insects, or chasing insects on muddy wetland edges with its head held low.

reddish brown markings on sides of back

grayish brown wings

♀ BREEDING

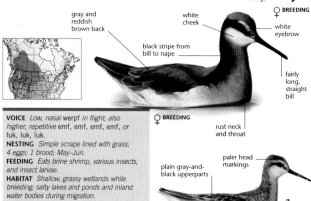

gray and reddish brown back

white cheek

black stripe from bill to nape

white eyebrow

fairly long, straight bill

♀ BREEDING

rust neck and throat

paler head markings

plain gray-and-black upperparts

♂

VOICE Low, nasal *werpf* in flight; also higher, repetitive *emf, emf, emf, emf*, or *luk, luk, luk*.
NESTING Simple scrape lined with grass; 4 eggs; 1 brood; May–Jun.
FEEDING Eats brine shrimp, various insects, and insect larvae.
HABITAT Shallow, grassy wetlands while breeding; salty lakes and ponds and inland water bodies during migration.
LENGTH 8½–9½in (22–24cm)
WINGSPAN 15½–17in (39–43cm)

Red-necked Phalarope

S

Phalaropus lobatus

This aquatic sandpiper spends nine months in deep ocean waters feeding on tiny plankton before coming to nest in the Arctic. Unlike most bird species, the female phalarope is more brightly colored and slightly larger than the male; after competing savagely for a male, she migrates right after laying her eggs and leaves the male to care for them.

pointed wings

narrow, white wing stripe

♀ BREEDING

needle-like, dark bill

dark gray crown and face

dark upperparts with buff or rust feather edges

♀ BREEDING

white throat

rust neck and upper breast

dark upperparts with buff stripes

VOICE Hard, squeaky *pwit* or *kit* in flight; variations of the same vocalizations on breeding grounds.
NESTING Depression in wet sedge or grass; 3–4 eggs; 1–2 broods; May–Jun.
FEEDING Plankton; also insects, brine shrimp, and mollusks.
HABITAT Wet tundra on raised ridges or hummocks while breeding; far out to sea during migration.
LENGTH 7–7½in (18–19cm)
WINGSPAN 12½–16in (32–41cm)

Red Phalarope

Phalaropus fulicarius

The Red Phalarope spends over ten months each year over deep ocean waters. It also migrates across the ocean; few birds of this species are ever seen inland. During migration over Alaskan waters, flocks feed on crustaceans in the mud plumes created by the foraging of gray and bowhead whales on the ocean floor.

broad, pointed wings

white rump with black line in center, and white edges

bold white wing bar

♀ BREEDING

black crown

tan-fringed feathers on upperparts

bold white cheek patch

stout, yellow bill with black tip

♀ BREEDING

deep brick-red neck, throat, and underparts

black cheek patch and nape

mostly gray upperparts

white neck and head

white underparts

NONBREEDING

VOICE *Sharp* psip *or* pseet *in flight, often in rapid succession; drawn-out, 2-syllabled* sweet *in alarm.*
NESTING *Depression on ridge or hummock in coastal sedge; 3–4 eggs; 1 brood; Jun.*
FEEDING *Marine crustaceans, fish eggs, larval fish; adult or larval insects.*
HABITAT *Coastal Arctic tundra while breeding; deep ocean waters during migration and winter.*
LENGTH *8–8½in (20–22cm)*
WINGSPAN *16–17½in (41–44cm)*

Heermann's Gull

Larus heermanni

In North America, the breeding Heermann's Gull is the only gull with a dark gray body and white head. These features, along with its bright red bill, make this gull unmistakable. Nonbreeding birds have a mottled dark head and a black-tipped bill. Juveniles are generally dark brown, with pale patches at the base of their bills. These gulls have black legs in all plumages.

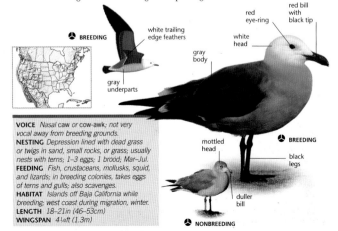

BREEDING

white trailing edge feathers

gray underparts

red eye-ring

red bill with black tip

white head

gray body

mottled head

BREEDING

black legs

duller bill

NONBREEDING

VOICE *Nasal* caw *or* cow-awk; *not very vocal away from breeding grounds.*
NESTING *Depression lined with dead grass or twigs in sand, small rocks, or grass; usually nests with terns; 1–3 eggs; 1 brood; Mar–Jul.*
FEEDING *Fish, crustaceans, mollusks, squid, and lizards; in breeding colonies, takes eggs of terns and gulls; also scavenges.*
HABITAT *Islands off Baja California while breeding; west coast during migration, winter.*
LENGTH *18–21in (46–53cm)*
WINGSPAN *4¼ft (1.3m)*

Ring-billed Gull

Ⓢ

Larus delawarensis

One of the most common birds in North America, the medium-sized Ring-billed Gull is distinguished by the black band on its yellow bill. From the mid-19th to the early 20th century, population numbers crashed due to egg and plumage hunting and habitat loss. Protection allowed the species to make a spectacular comeback, and in the 1990s, there were an estimated 3–4 million birds. It can often be seen scavenging in parking lots at malls and picnic sites.

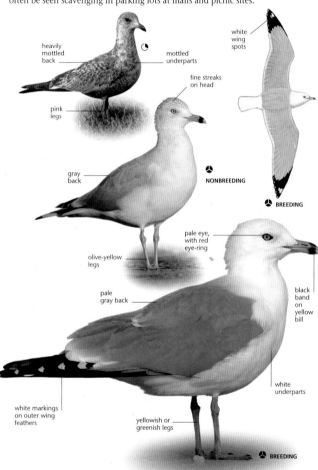

heavily
mottled
back

mottled
underparts

fine streaks
on head

white
wing
spots

pink
legs

gray
back

NONBREEDING

BREEDING

pale eye,
with red
eye-ring

olive-yellow
legs

pale
gray back

black
band
on
yellow
bill

white
underparts

white markings
on outer wing
feathers

yellowish or
greenish legs

BREEDING

VOICE *Slightly nasal, whiny kee-ow or meee-ow; series of 4–6
kyaw notes, higher pitched than Herring Gull.*
NESTING *Shallow cup of plant matter on ground in open areas,
usually near low vegetation; 1–5 eggs; 1 brood; Apr–Aug.*
FEEDING *Picks food while walking or dips and plunges in water;
small fish, insects, grain, small rodents; also scavenges.*
HABITAT *Freshwater habitats while breeding; mostly saltwater
areas near coasts, and along major river systems in winter.*
LENGTH *17–21½in (43–54cm)*
WINGSPAN *4–5ft (1.2–1.5m)*

Mew Gull

Larus canus

The Mew Gull's small bill and rounded head give it a rather dove-like profile. It can be confused with the widespread Ring-billed Gull, which it resembles in all plumages. Some taxonomists split the Mew Gull into four species—the European "Common Gull" (*L. c. canus*), the northeast Asian species (*L. c. heinei*), the "Kamchatka Gull" (*L. c. kamtschatschensis*), and the North American "Short-billed Gull" (*L. c. brachyrhynchus*).

white spot on wing tip

🦅 BREEDING

small bill, often with dusky ring

streaks on rounded head

all-yellow bill

dark gray back

dusky mottling

yellow legs

🦅 BREEDING

yellow to green legs

🦅 NONBREEDING

VOICE *Shrill mewing calls; higher pitched than other gulls.*
NESTING *Platform of vegetation in trees or on ground; 1–5 eggs; 1 brood; May–Aug.*
FEEDING *Aquatic crustaceans and mollusks, insects, fish, bird eggs, chicks; scavenges trash and steals food from other birds.*
HABITAT *Tundra, marshy areas, lakes and rivers, and coastal cliffs while breeding; coastal waters, beaches, and mudflats during winter.*
LENGTH *15–16in (38–41cm)*
WINGSPAN *3ft 3in–4ft (1–1.2m)*

California Gull

Larus californicus

Slightly smaller than the Herring Gull, the California Gull has a darker back and longer wings. In breeding plumage, it can also be distinguished by the black and red coloration on its bill and its greenish yellow legs. In winter and on young birds, dark streaks are prominent on the nape of the neck.

black wing tips with white terminal spot

🦅 NONBREEDING

dark streaks on nape of neck

red eye-ring

black line and red spot on bill

white head and neck

gray back

white trailing edge to feathers

black-tipped wings

white underparts

🦅 NONBREEDING

🦅 BREEDING

greenish yellow legs and toes

VOICE *Call a repeated kee-yah, kee-yah, kee-yah.*
NESTING *Shallow scrape lined with feathers, bones, and vegetation, usually on islands; 2–3 eggs; 1 brood; May–Jul.*
FEEDING *Forages around lakes for insects, mollusks; hovers over cherry trees dislodging fruits with its wings.*
HABITAT *Islands on lakes and rivers in inland plains while breeding; Pacific coast in winter.*
LENGTH *17½–20in (45–51cm)*
WINGSPAN *4–4½ft (1.2–1.4m)*

Great Black-backed Gull ⓢ

Larus marinus

The largest gull in North America, the Great Black-backed Gull has a bullying disposition. In breeding colonies, after their eggs hatch, adults dive at ground predators and strike them with their wings and feet. Terns and eiders that nest near aggressive Great Black-backed Gull colonies may suffer low rates of nest predation, but their newly hatched fledglings are often swallowed whole by the gulls.

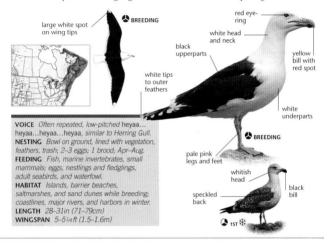

large white spot on wing tips

🐦 BREEDING

red eye-ring

white head and neck

black upperparts

yellow bill with red spot

white tips to outer feathers

white underparts

🐦 BREEDING

pale pink legs and feet

whitish head

black bill

speckled back

🕐 1ST ❄

VOICE *Often repeated, low-pitched heyaa... heyaa...heyaa...heyaa, similar to Herring Gull.*
NESTING *Bowl on ground, lined with vegetation; feathers, trash; 2–3 eggs; 1 brood; Apr–Aug.*
FEEDING *Fish, marine invertebrates, small mammals; eggs, nestlings and fledglings, adult seabirds, and waterfowl.*
HABITAT *Islands, barrier beaches, saltmarshes, and sand dunes while breeding; coastlines, major rivers, and harbors in winter.*
LENGTH *28–31in (71–79cm)*
WINGSPAN *5–5¼ft (1.5–1.6m)*

Glaucous-winged Gull ⓢ

Larus glaucescens

The Glaucous-winged Gull, the most common large gull on the north Pacific coast, is found in cities, even nesting on the roofs of shorefront buildings. This species commonly interbreeds with Western Gulls in the southern part of its range, and with Herring and Glaucous Gulls in the north, producing intermediate birds that are more difficult to identify.

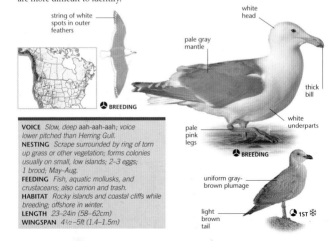

string of white spots in outer feathers

🐦 BREEDING

white head

pale gray mantle

thick bill

white underparts

pale pink legs

🐦 BREEDING

uniform gray-brown plumage

light brown tail

🕐 1ST ❄

VOICE *Slow, deep aah-aah-aah; voice lower pitched than Herring Gull.*
NESTING *Scrape surrounded by ring of torn up grass or other vegetation; forms colonies usually on small, low islands; 2–3 eggs; 1 brood; May–Aug.*
FEEDING *Fish, aquatic mollusks, and crustaceans; also carrion and trash.*
HABITAT *Rocky islands and coastal cliffs while breeding; offshore in winter.*
LENGTH *23–24in (58–62cm)*
WINGSPAN *4½–5ft (1.4–1.5m)*

Western Gull

Larus occidentalis

The Western Gull is the only dark-backed gull found regularly within its normal range and habitat. Identification is complicated due to two subspecies: the paler *occidentalis* in the north, and the darker *wymani* in the south. Western Gulls interbreed with Glaucous-winged Gulls, producing confusing hybrids. Occasionally, two females will each lay eggs and then raise their broods together.

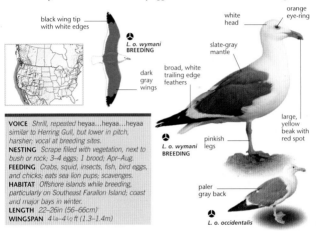

black wing tip with white edges

L. o. wymani BREEDING

dark gray wings

broad, white trailing edge feathers

white head

orange eye-ring

slate-gray mantle

large, yellow beak with red spot

L. o. wymani BREEDING

pinkish legs

paler gray back

L. o. occidentalis

VOICE *Shrill, repeated heyaa...heyaa...heyaa similar to Herring Gull, but lower in pitch, harsher; vocal at breeding sites.*
NESTING *Scrape filled with vegetation, next to bush or rock; 3–4 eggs; 1 brood; Apr–Aug.*
FEEDING *Crabs, squid, insects, fish, bird eggs, and chicks; eats sea lion pups; scavenges.*
HABITAT *Offshore islands while breeding, particularly on Southeast Farallon Island; coast and major bays in winter.*
LENGTH *22–26in (56–66cm)*
WINGSPAN *4¼–4½ft (1.3–1.4m)*

Glaucous Gull

Larus hyperboreus

The Glaucous Gull is the largest of the "white-winged" gulls; it appears like a large white spectre among its smaller, darker cousins. In the Arctic, successful pairs of Glaucous Gulls maintain pair bonds with their mates for years, often returning to the same nest site year after year.

streaking on head

yellow bill with distinct red spot

white head

pale gray upperparts

white underparts

white wing tips

pink legs

white underparts

mottled, pale brown back

pale brown underparts

1ST

VOICE *Similar to the Herring Gull, but slightly harsher and deeper; hoarse, nasal ku-ku-ku.*
NESTING *Shallow cup lined with vegetation on ground, at edge of tundra pools, on cliffs and islands; 1–3 eggs; 1 brood; May–Jul.*
FEEDING *Fish, crustaceans, mollusks; also eggs and chicks of waterfowl, small seabirds, and small mammals.*
HABITAT *High-Arctic coast while breeding; coastlines and Great Lakes in winter.*
LENGTH *26–30in (65–75cm)*
WINGSPAN *5–6ft (1.5–1.8m)*

Iceland Gull

Ⓢ

Larus glaucoides

The Iceland Gull is the smallest "white-winged" gull. It is more often seen in winter, and immatures more than adults. North American breeding birds have gray wing tips and have been considered a separate species called the "Kumlien's Gull." The subspecies *L. g. glaucoides* possesses white wing tips; it is found mostly in Greenland and Iceland, but a few birds travel to the western North Atlantic.

gray wing tips

markedly streaked head

gray back

pale or gray wing tip

short, pale yellow bill with red spot

pink legs

white belly

brown barred plumage

blackish bill

Ⓛ 1ST ❄

VOICE Clew, clew, clew or kak-kak-kak; virtually silent on wintering grounds.
NESTING Loose nest of moss, vegetation, and feathers, usually on narrow rock ledge; 2–3 eggs; 1 brood; May–Aug.
FEEDING Small fish while in flight; crustaceans, mollusks, carrion, and garbage.
HABITAT Ledges on vertical sea cliffs while breeding; open water along coastline and occasionally Great Lakes in winter.
LENGTH 20½–23½in (52–60cm)
WINGSPAN 4½–5ft (1.4–1.5m)

Thayer's Gull

Ⓢ

Larus thayeri

The classification of Thayer's Gull is controversial; it was originally considered a subspecies of the Herring Gull, then a full species, and now many authorities consider the Thayer's Gull to be a subspecies of the Iceland Gull. When standing with the Herring and Iceland Gulls, this bird is difficult to identify, a process complicated further by the existence of hybrid gulls of various parentages.

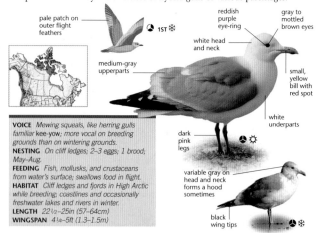

pale patch on outer flight feathers

Ⓛ 1ST ❄

reddish purple eye-ring

gray to mottled brown eyes

white head and neck

medium-gray upperparts

small, yellow bill with red spot

white underparts

dark pink legs

☀

variable gray on head and neck forms a hood sometimes

black wing tips

Ⓐ ❄

VOICE Mewing squeals, like herring gulls familiar kee-yow; more vocal on breeding grounds than on wintering grounds.
NESTING On cliff ledges; 2–3 eggs; 1 brood; May–Aug.
FEEDING Fish, mollusks, and crustaceans from water's surface; swallows food in flight.
HABITAT Cliff ledges and fjords in High Arctic while breeding; coastlines and occasionally freshwater lakes and rivers in winter.
LENGTH 22½–25in (57–64cm)
WINGSPAN 4¼–5ft (1.3–1.5m)

Herring Gull

S

Larus argentatus

The Herring Gull is the archetypal, large "white-headed" gull that nearly all other gulls are compared with. When people mention "seagulls" they usually refer to this species, however the Herring Gull, like most other gulls, does not commonly go far out to sea—it is a bird of near-shore waters, coasts, lakes, rivers, and inland waterways. Now more common, the Herring Gull was almost wiped out in the late 19th and early 20th century by plumage hunters and egg collectors.

streaked head and neck

NONBREEDING

white spots near wing tips

gray wings

BREEDING

mottled brown back

barred brown body

1ST

white head and neck

gray back

large, yellow bill with red spot

white underparts

black outer wing feathers

pink legs

BREEDING

VOICE *High-pitched, shrill, repeated heyaa...heyaa...heyaa...heyaa.*
NESTING *Shallow bowl on ground lined with feathers, vegetation, detritus; 2–4 eggs; 1 brood; Apr–Aug.*
FEEDING *Fish, crustaceans, mollusks, worms; eggs and chicks of other seabirds; scavenges carrion, garbage; steals from other birds.*
HABITAT *Coasts, and inland on lakes, rivers, reservoirs, and garbage dumps.*
LENGTH *22–26in (56–66cm)*
WINGSPAN *4–5ft (1.2–1.5m)*

Lesser Black-backed Gull

S

Larus fuscus

This bird has become an annual winter visitor to the east coast of North America. Nearly all the Lesser Black-backed Gulls found in North America are of the Icelandic and western European subspecies *L. f. graellsii*, with a slate-gray back. Another European subspecies, with a much darker back, has rarely been reported in North America.

streaked head and neck

black wing tips with white spot

slate-gray back

NONBREEDING

yellow eye

white underparts

NONBREEDING

dull yellow legs

white head

BREEDING

yellow bill with red spot

bright yellow legs

VOICE Kyow...yow...yow...yow, *similar to Herring Gull; also a deeper and throaty, repeated gah-gah-gah-gah.*
NESTING *Scrape on ground lined with lichens, grass, and feathers; 3 eggs; 1 brood; Apr–Sep.*
FEEDING *Mollusks, crustaceans, and various insects; also scavenges carrion and garbage.*
HABITAT *Winter visitor to eastern coast, at harbors and near fishing boats, and inland at lakeshores and landfills.*
LENGTH 20½–26in (52–67cm)
WINGSPAN 4¼–5ft (1.3–1.5m)

Black-headed Gull

S

Chroicocephalus ridibundus

An abundant breeder in Eurasia, the Black-headed Gull colonized North America in the 20th century. It has become common in Newfoundland after being found nesting there in 1977, and has nested as far south as Cape Cod. However, it has not spread far to the West and remains an infrequent visitor or vagrant over most of the continent.

white flash on outer wings

black-tipped red bill

NONBREEDING

black trailing edge of wing

chocolate brown hood

white nape

very pale gray back

dark red bill

brownish "crown-collar"

reddish bill

gray back

dark red legs

BREEDING

dark "ear" spot

white underparts

bright red legs

VOICE *Loud laughing or a chattering kek kek keeaar; vocal at breeding sites.*
NESTING *Loose mass of vegetation, on ground or on top of other vegetation; 2–3 eggs; 1 brood; Apr–Aug.*
FEEDING *Insects, small crustaceans, and mollusks; some vegetation; forages in plowed farm fields; raids garbage dumps.*
HABITAT *Harbors, inlets, bays, rivers, lakes, and garbage dumps.*
LENGTH 13½–14½in (34–37cm)
WINGSPAN 3ft 3in–3½ft (1–1.1m)

NONBREEDING

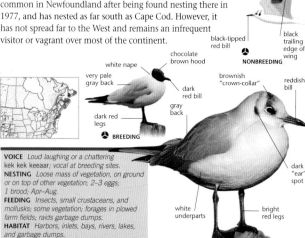

Bonaparte's Gull

Chroicocephalus philadelphia

Lighter and more delicate than the other North American gulls, Bonaparte's Gull is commonly distinguished in winter by the blackish smudge behind each eye and the large, white wing patch. It is one of North America's most common and widespread gulls. This species was named after the 19th century French ornithologist Charles Lucien Bonaparte (nephew of Napoleon).

black wing tips

white flash on outer wings

NONBREEDING

white head

blackish "ear" spot

gray neck

NONBREEDING

short bill

black hood

gray back and wings

white wedge on wing

orange-red legs

white underparts with rosy glow when breeding

BREEDING

VOICE Harsh keek, keek; kew, kew, kew while feeding.
NESTING Stick nest of twigs, bark, lined with mosses or lichens; in conifers or rushes over water; 1–4 eggs; 1 brood; May–Jul.
FEEDING Insects; crustaceans, mollusks, small fish from water's surface; also plunge-dives.
HABITAT Northern forests in lakes, ponds, or bogs while breeding; near water during migration; Great Lakes, coasts in winter.
LENGTH 11–12in (28–30cm)
WINGSPAN 35in–3ft 3in (90–100cm)

Laughing Gull

Leucophaeus atricilla

The distinctive call of the Laughing Gull is a familiar sound in spring and summer along the East Coast. Greatly reduced in the 19th century by egg and plumage hunting, its numbers increased in the 1920s following protection. As an irregular breeder in Canada, its populations have changed little since the 1970s.

dark gray wings

gray nape

black head

long, slightly drooped bill

broken white eye-ring

white neck

dark gray back

black wing tips

BREEDING

white underparts

long, dark legs

VOICE Strident laugh, ha...ha...ha...ha...ha; vocal in breeding season; quiet in winter.
NESTING Mass of grass on dry land with heavy vegetation, sand, rocks, and salt marshes; 2–4 eggs; 1 brood; Apr–Jul.
FEEDING Insects, worms, squid, crabs, crab eggs, larvae; also small fish, garbage, berries.
HABITAT Near saltwater while breeding; coasts, bays, estuaries, and lakes during migration and winter.
LENGTH 15½–18in (39–46cm)
WINGSPAN 3¼–4ft (1–1.2m)

Franklin's Gull

S

Leucophaeus pipixcan

Franklin's Gull is named after British Arctic explorer, John Franklin, on whose first expedition the bird was discovered in 1823. Unlike other gulls, this species has two complete molts each year. As a result, its plumage usually looks fresh and it rarely has the scruffy look of some other gulls.

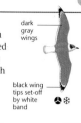

dark gray wings

black wing tips set-off by white band

☀❄

white in outer wing feathers

dark gray back

broken white eye crescent

black head

red bill

pink blush underneath

☀☀

dark back of head

gray back

short, straight bill

☀❄-

VOICE *Nasal weeh-a, weeh-a; shrill kuk kuk kuk kuk; extremely vocal on breeding colonies.*
NESTING *Floating mass of bulrushes or other plants, refreshed as nest sinks; 2–4 eggs; 1 brood; Apr–Jul.*
FEEDING *Earthworms, insects, and some seeds during breeding; opportunistic feeder during migration and winter.*
HABITAT *High prairies near water while breeding; agricultural areas during migration.*
LENGTH *12½–14in (32–36cm)*
WINGSPAN *33in–3ft 1in (85–95cm)*

Little Gull

S

Hydrocoloeus minutus

A Eurasian species, the Little Gull is the smallest gull in the world. It was first recorded in North America in the early1800s, but a nest was not found until 1962, in Ontario. Known nesting areas are still few, but winter numbers have been increasing steadily in recent decades.

black underwings

NONBREEDING

pale wing tips

pale gray back

pale head, with dark markings

thin, dark bill

NONBREEDING

black hood and bill

BREEDING

red legs

VOICE *Nasal kek, kek, kek, kek, reminiscent of a small tern.*
NESTING *Thick, floating mass of dry cattails, reeds, or other vegetation, in marshes and ponds; 3 eggs; 1 brood; May–Aug.*
FEEDING *Flying insects, aquatic invertebrates such as shrimps, and small fish.*
HABITAT *Extensive freshwater marshes while breeding; sea coasts and sewage outfalls in winter.*
LENGTH *10–12in (25–30cm)*
WINGSPAN *23½–26in (60–65cm)*

Sabine's Gull Ⓢ

Xema sabini

This gull was discovered in Greenland by the English scientist,
Edward Sabine, during John Ross's search for the Northwest Passage
in 1818. The distinctive wing pattern and notched tail make it
unmistakable in all plumages. This species breeds in the Arctic and
winters at sea, off the coasts of the Americas and Africa.

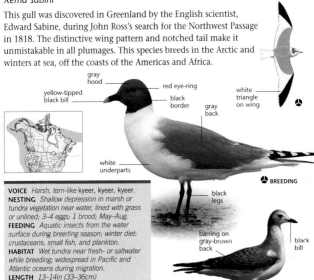

gray
hood

yellow-tipped
black bill

red eye-ring

black
border

gray
back

white
triangle
on wing

white
underparts

gray
back

BREEDING

black
legs

barring on
gray-brown
back

black
bill

VOICE *Harsh, tern-like kyeer, kyeer, kyeer.*
NESTING *Shallow depression in marsh or
tundra vegetation near water, lined with grass
or unlined; 3–4 eggs; 1 brood; May–Aug.*
FEEDING *Aquatic insects from the water
surface during breeding season; winter diet:
crustaceans, small fish, and plankton.*
HABITAT *Wet tundra near fresh- or saltwater
while breeding; widespread in Pacific and
Atlantic oceans during migration.*
LENGTH *13–14in (33–36cm)*
WINGSPAN *35in–3ft 3in (90–100cm)*

Black-legged Kittiwake Ⓢ

Rissa tridactyla

A kittiwake nesting colony may have thousands of birds lined up along steep,
narrow cliff ledges by the sea. Kittiwakes have sharper claws than other gulls,
probably to give them a better grip on their ledges. In the late 20th century,
the Black-legged Kittiwake population expanded greatly in the Canadian
maritime provinces, with numbers doubling in the Gulf of St. Lawrence.

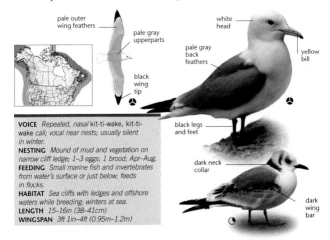

pale outer
wing feathers

pale gray
upperparts

white
head

pale gray
back
feathers

yellow
bill

black
wing
tip

black legs
and feet

dark neck
collar

dark
wing
bar

VOICE *Repeated, nasal kit-ti-wake, kit-ti-
wake call; vocal near nests; usually silent
in winter.*
NESTING *Mound of mud and vegetation on
narrow cliff ledge; 1–3 eggs; 1 brood; Apr–Aug.*
FEEDING *Small marine fish and invertebrates
from water's surface or just below; feeds
in flocks.*
HABITAT *Sea cliffs with ledges and offshore
waters while breeding; winters at sea.*
LENGTH *15–16in (38–41cm)*
WINGSPAN *3ft 1in–4ft (0.95m–1.2m)*

Caspian Tern (S)

Hydroprogne caspia

The Caspian Tern is the world's largest tern. Unlike other "black-capped" terns, it never has a completely white forehead, even in winter. It steals prey from other seabirds, and also snatches eggs from, and hunts the nestlings of, other gulls and terns. It aggressively defends its nesting territory, giving hoarse alarm calls, and rhythmically opening and closing its beak in a threatening display.

VOICE *Hoarse, deep kraaa, kraaa; also barks at intruders; male's wings vibrate loudly in courtship flight.*
NESTING *Shallow scrape on ground; 2–3 eggs; 1 brood; May–Aug.*
FEEDING *Plunges into water to snatch fish, barnacles, and snails.*
HABITAT *Interior lakes, saltmarshes, and coastal barrier islands while breeding; wetlands during migration; coasts in winter.*
LENGTH *18½–21½in (47–54cm)*
WINGSPAN *4¼–5ft (1.3–1.5m)*

Roseate Tern ●

Sterna dougallii

Mostly found nesting with the Common Tern, the Roseate Tern is paler and more slender. Its bill is black for a short time in the spring before turning at least half red during the nesting season. Pairs glide down from hundreds of feet in the air in courtship flights, swaying side to side with each other. Two females and a male may nest together, sharing egg incubation and rearing of young.

VOICE *Keek or ki-rik in flight and around nesting colony.*
NESTING *Simple scrape, often under vegetation or large rocks; adds twigs and dry grass during incubation; 1–3 eggs; 1 brood; May–Aug.*
FEEDING *Catches small fish in bill by diving from the air; carries whole fish to young.*
HABITAT *Coastal beaches and offshore islands while breeding.*
LENGTH *13–16in (33–41cm)*
WINGSPAN *28in (70cm)*

Common Tern

Sterna hirundo

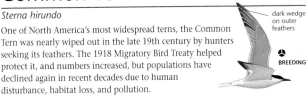

dark wedge
on outer
feathers

BREEDING

One of North America's most widespread terns, the Common
Tern was nearly wiped out in the late 19th century by hunters
seeking its feathers. The 1918 Migratory Bird Treaty helped
protect it, and numbers increased, but populations have
declined again in recent decades due to human
disturbance, habitat loss, and pollution.

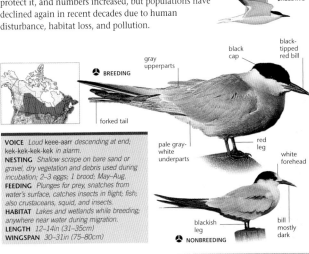

BREEDING

gray
upperparts

black
cap

black-
tipped
red bill

forked tail

pale gray-
white
underparts

red
leg

white
forehead

blackish
leg

bill
mostly
dark

NONBREEDING

VOICE *Loud keee-aarr descending at end;
kek-kek-kek-kek in alarm.*
NESTING *Shallow scrape on bare sand or
gravel, dry vegetation and debris used during
incubation; 2–3 eggs; 1 brood; May–Aug.*
FEEDING *Plunges for prey, snatches from
water's surface, catches insects in flight; fish;
also crustaceans, squid, and insects.*
HABITAT *Lakes and wetlands while breeding;
anywhere near water during migration.*
LENGTH *12–14in (31–35cm)*
WINGSPAN *30–31in (75–80cm)*

Arctic Tern

Sterna paradisaea

Most Arctic Terns breed in the Arctic, then migrate to the Antarctic seas for the
Southern Hemisphere summer before returning north, a round trip of at least
25,000 miles (40,000km). Apart from during migration, it spends its life in areas of
near continuous daylight and rarely comes to land, except to nest. It can be
distinguished from the Common Tern by its smaller bill, and shorter legs and neck.

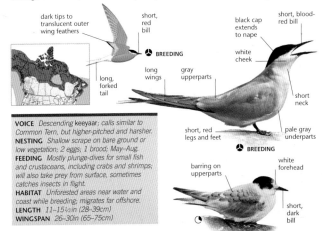

dark tips to
translucent outer
wing feathers

short,
red
bill

BREEDING

black cap
extends
to nape

short, blood-
red bill

white
cheek

gray
upperparts

long
wings

long,
forked
tail

short
neck

pale gray
underparts

short, red
legs and feet

BREEDING

barring on
upperparts

white
forehead

short,
dark
bill

VOICE *Descending keeyaar; calls similar to
Common Tern, but higher-pitched and harsher.*
NESTING *Shallow scrape on bare ground or
low vegetation; 2 eggs; 1 brood; May–Aug.*
FEEDING *Mostly plunge-dives for small fish
and crustaceans, including crabs and shrimps;
will also take prey from surface, sometimes
catches insects in flight.*
HABITAT *Unforested areas near water and
coast while breeding; migrates far offshore.*
LENGTH *11–15½in (28–39cm)*
WINGSPAN *26–30in (65–75cm)*

Forster's Tern

S

Sterna forsteri

This medium-sized tern can be differentiated from the Common Tern by its lighter outer wing feathers and longer tail. Early naturalists could not tell the two species apart until 1834 when English botanist Thomas Nuttall made the distinction. He named this tern after Johann Reinhold Forster, a naturalist who accompanied the English explorer Captain Cook on his second voyage (1772-75).

deeply forked tail

gray wings with slightly darker wing tips

NONBREEDING

pale gray upperparts

black cap and nape

orange-red bill with dark tip

BREEDING

long, gray tail with white outer margins

snowy white underparts

plain gray wings

dark bill

NONBREEDING

VOICE *Harsh, descending kyerr; more nasal than Common Tern.*
NESTING *Shallow scrape in mud or sand; occasionally nests on top of muskrat lodge or on old grebe nest, or raft of floating vegetation; 2–3 eggs; 1 brood; May–Aug.*
FEEDING *Plunge-dives for fish and crustaceans; also catches insects in flight.*
HABITAT *Fresh- and saltwater marshes with open water while breeding; coasts in winter.*
LENGTH *13–14in (33–36cm)*
WINGSPAN *29–32in (73–82cm)*

Black Tern

S

Chlidonias niger

This marsh-dwelling tern undergoes a remarkable change in appearance from summer to winter. The Black Tern's breeding plumage resembles the closely related White-winged Tern, an accidental visitor to North America. The Black Tern's nonbreeding plumage is much paler than its breeding plumage—the head turns white with irregular black streaks, and the neck, breast, and belly become whitish gray.

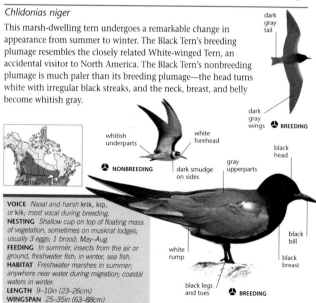

dark gray tail

dark gray wings

BREEDING

whitish underparts

white forehead

NONBREEDING

dark smudge on sides

gray upperparts

black head

black bill

black breast

white rump

black legs and toes

BREEDING

VOICE *Nasal and harsh krik, kip, or kik; most vocal during breeding.*
NESTING *Shallow cup on top of floating mass of vegetation, sometimes on muskrat lodges; usually 3 eggs; 1 brood; May–Aug.*
FEEDING *In summer, insects from the air or ground, freshwater fish; in winter, sea fish.*
HABITAT *Freshwater marshes in summer; anywhere near water during migration; coastal waters in winter.*
LENGTH *9–10in (23–26cm)*
WINGSPAN *25–35in (63–88cm)*

Pomarine Jaeger

Stercorarius pomarinus

The Pomarine Jaeger uses its size and strength to overpower larger seabirds, such as gulls and shearwaters, in order to steal their food. Nesting only when populations of lemmings are at their peak to provide food for its young, it is readily driven away from breeding territories by the more dynamic Parasitic Jaeger. Research suggests that the Pomarine Jaeger is actually more closely related to the large skuas than to other jaegers.

prominent white "flash" in feathers

BREEDING PALE FORM

white wing flash

DARK FORM

blunt tail spike

dark overall

NONBREEDING PALE FORM

blackish cap

pale based, thick bill

cream cheeks

gray-brown back

barred flanks

dusky breastband

twisted, spoon-like central tail feathers

dusky breast-band

BREEDING PALE FORM

VOICE *Nasal cow-cow-cow and various sharp, low whistles.*
NESTING *Shallow unlined depression on a rise or hummock in open tundra; 2 eggs; 1 brood; Jun–Aug.*
FEEDING *Lemmings and other rodents; fish; scavenges refuse from fishing boats during nonbreeding season; steals fish from other seabirds.*
HABITAT *Open tundra while breeding; coasts and far offshore during migration; more common on West Coast.*
LENGTH *17–20in (43–51cm)*
WINGSPAN *4ft (1.2m)*

Parasitic Jaeger Ⓢ

Stercorarius parasiticus

The Parasitic Jaeger routinely seeks food by chasing, bullying, and forcing other seabirds to drop or regurgitate fish or other food they have caught. Unlike most jaegers, the Parasitic Jaeger is adaptable in its feeding habits so that it can forage and raise its young under a wide range of environmental conditions. Breeding on the Arctic tundra, it migrates to offshore areas during the nonbreeding season.

white wing patch

🦅 PALE FORM

pale cheek

dark cap

gray breastband

long, pointed, central feathers

dark upperparts

🦅 PALE FORM

dark legs and toes

white wing patch

pale cheek patch

mostly dark brown overall

🦅 DARK FORM

VOICE *Terrier-like yelps and soft squeals, often during interactions with other jaegers or predators, usually around nesting territories.*
NESTING *Shallow unlined depression on a rise in open tundra; 2 eggs; 1 brood; May–Aug.*
FEEDING *Steals fish and other aquatic prey from gulls and terns; small birds, eggs, or small rodents on breeding grounds.*
HABITAT *Tundra while breeding; near- and offshore waters during migration and winter.*
LENGTH *16–18½in (41–47cm)*
WINGSPAN *3ft 3in–3½ft (1–1.1m)*

Long-tailed Jaeger Ⓢ

Stercorarius longicaudus

This elegant species is a surprisingly fierce Arctic and marine predator. The Long-tailed Jaeger occasionally steals food from small gulls and terns, but usually hunts for its own food. In years when lemming numbers dip so low as to become unavailable for feeding its nestlings, this jaeger may not even attempt to nest.

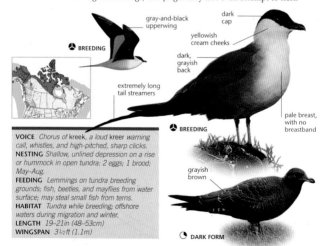

gray-and-black upperwing

dark cap

🦅 BREEDING

yellowish cream cheeks

dark, grayish back

extremely long tail streamers

🦅 BREEDING

pale breast, with no breastband

grayish brown

🌓 DARK FORM

VOICE *Chorus of* kreek, *a loud* kreer *warning call, whistles, and high-pitched, sharp clicks.*
NESTING *Shallow, unlined depression on a rise or hummock in open tundra; 2 eggs; 1 brood; May–Aug.*
FEEDING *Lemmings on tundra breeding grounds; fish, beetles, and mayflies from water surface; may steal small fish from terns.*
HABITAT *Tundra while breeding; offshore waters during migration and winter.*
LENGTH *19–21in (48–53cm)*
WINGSPAN *3½ft (1.1m)*

Dovekie

(S)

Alle alle

Also known as the Little Auk, the Dovekie is a bird of the High Arctic. Most Dovekies breed in Greenland in large, noisy, crowded colonies, but some breed in northeastern Canada, and others on a few islands in the Bering Sea off Alaska. Vast flocks of Dovekies winter on the Low Arctic waters off the northeastern North American seaboard. Severe onshore gales may cause entire flocks to become stranded along the East Coast of North America.

short, dark tail

dark wings

🌓 BREEDING

dark back

white collar at back of head

dark crown

small bill

white throat

white undertail

🌓 NONBREEDING

dark head and upper breast

white triangle on side of breast

🌓 BREEDING

VOICE High-pitched trilling that rises and falls at breeding ground; silent at sea.
NESTING Pebble nest in crack or crevice in boulder field or rocky outcrop; 1 egg; 1 brood; Apr–Aug.
FEEDING Tiny crustaceans from just below the sea's surface.
HABITAT Islands while breeding; just south of Arctic ice pack and northeastern seaboard in winter.
LENGTH 8½in (21cm)
WINGSPAN 15in (38cm)

Thick-billed Murre

(S)

Uria lomvia

The Thick-billed Murre is one of the most abundant seabirds in the whole of the Northern Hemisphere, breeding in dense, coastal cliff colonies of around a million birds each. Young leave the colony when they are only about 25 percent of the adult's weight, completing their growth at sea while being fed by the male parent alone. This murre can dive to 600ft (180m) to catch fish and squid.

short, black tail

🌓 BREEDING

hunched in flight

brownish black sides of head

more extensive white on throat

reduced or absent white line on bill

🌓 NONBREEDING

white line along bill

white breast and underparts

all-blackish upperparts

VOICE Roaring, groaning, insistent sounding aoorrr; lower-pitched than the Common Murre.
NESTING Rocky coast or narrow sea cliff ledge in dense colony; 1 egg; 1 brood; Mar–Sep.
FEEDING Cod, herring, capelin, and sand lance in summer; also crustaceans, worms, and squid.
HABITAT Rocky shorelines while breeding; winters at sea, often near pack ice edges or openings.
LENGTH 18in (46cm)
WINGSPAN 28in (70cm)

🌓 BREEDING

Common Murre

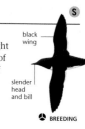

black wing

slender head and bill

BREEDING

Uria aalge

Penguin-like Common Murres are often seen standing upright on cliffs. They are strong fliers and adept divers, to a depth of 500ft (150m). Their large nesting colonies, on rocky sea cliff ledges, are so crowded that incubating adults may actually touch each other on both sides. Common Murre eggs are pointed at one end—when pushed, they roll in a circle, reducing the risk of rolling off the nesting ledge.

curved, black line droops behind eye

white face and throat

NONBREEDING

black head

long, straight, black bill

white underparts

black back

grayish legs and feet

BREEDING

VOICE Low-pitched, descending call, reminiscent of trumpeting elephant.
NESTING On bare rock near shore, on cliff ledge, or in crevice; 1 egg; 1 brood; May–Jul.
FEEDING Small schooling fish, such as herring, sand lance, and haddock; also crustaceans, marine worms, and squid.
HABITAT Close to rocky shorelines, on cliff ledges or on top of sea stacks while breeding; winters at sea.
LENGTH 17½in (44cm)
WINGSPAN 26in (65cm)

Razorbill

thick, black bill

long, black, pointed tail

BREEDING

Alca torda

The Razorbill is the closest living relative of the extinct Great Auk. One of the rarest breeding seabirds in North America, it is a strong and agile flier. Razorbills typically feed at depths of about 20ft (6m), but can dive to depths of more than 450ft (140m). On shore, Razorbills walk upright like penguins. They carry small fish to their young; later, male razorbills escort their flightless young to the sea to feed.

bill smaller than in breeding birds

brownish head

white underparts up to chin

NONBREEDING

large, round head

thin white line extends from bill to eye

short neck

black upp3rparts

snowy white underparts

blackish legs and feet

BREEDING

VOICE Deep, guttural, resonant croak, hey al.
NESTING Enclosed sites often built in crevices, among boulders, or in abandoned burrows; 1 egg; 1 brood; May–Jul.
FEEDING Schooling fish (capelin, herring, sand lance); also marine worms and crustaceans; sometimes steals fish from other auks.
HABITAT Rocky islands and shorelines, or steep mainland cliffs while breeding; ice-free coastal waters in winter.
LENGTH 17in (43cm)
WINGSPAN 26in (65cm)

Black Guillemot

Cepphus grylle

Black Guillemots, also known as "sea pigeons," are medium-sized auks with distinctive black plumage and white wing patches. Their striking scarlet legs and mouth lining help attract a mate during the breeding season. Black Guillemots prefer shallow, inshore waters to the open ocean. They winter near the shore, sometimes moving into the mouths of rivers.

broad, rounded wings

oval, snowy white upperwing patch

🦅 BREEDING

large white patch

thin, straight bill

round, black body

dark belly

🦅 BREEDING

scarlet legs and feet

gray bars in white wing patch

gray cap

gray neck

VOICE *High-pitched whistles and squeaks near nesting habitat that resonate like an echo.*
NESTING *Shallow scrape in soil or pebbles within cave or crevice; 1–2 eggs; 1 brood; May–Aug.*
FEEDING *Small, bottom-dwelling fish, such as rock eels, sand lance, and sculpin; feeds close to nesting islands.*
HABITAT *Remote rocky islands and cliffs while breeding; shallow waters near rocky coasts.*
LENGTH *13in (33cm)*
WINGSPAN *21in (53cm)*

Pigeon Guillemot

Cepphus columba

The Pigeon Guillemot, a North Pacific seabird, is found along rocky shores in small colonies or isolated pairs. This auk nests in burrows or under rocks, often on small islands that provide protection from land-bound predators. The male excavates or chooses an abandoned burrow to build a nest. The bird's red-orange legs and mouth lining are used in courtship displays.

feet and legs trail in flight

oval, snowy white upperwing patch

dusky crown

dusky neck and face

black upperparts

stocky, round body

broad, rounded wings

dark rump

dark bar across white wing patch

bright red-orange legs and feet

🦅 BREEDING

VOICE *Excited, squeaky whistles, and twitters; weak whistle peeeee at nest.*
NESTING *Shallow scrape in burrow or crevice; 2 eggs; 1 brood; May–Aug.*
FEEDING *Feeds near shore; dives for rock eels, sculpin, crabs, shrimp, marine worms, and mollusks; carries food for young in beak.*
HABITAT *Rocky islands, coasts, and cliffs while breeding; waters off rocky coasts, and edges of pack ice in Bering Sea.*
LENGTH *13½in (34cm)*
WINGSPAN *23in (58cm)*

Marbled Murrelet

Brachyramphus marmoratus

The Marbled Murrelet feeds at sea and nests sometimes high up in trees in old growth forests. Unlike most auks and their relatives, which have black and white breeding plumage, its breeding plumage is brown to camouflage the bird on its nest in the branches of trees or, in places, on the ground. Its numbers are declining due to clear-cutting of old-growth forests, oil pollution and fishing net entanglement.

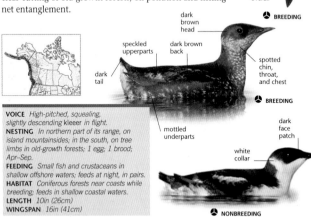

dark overall

🟤 BREEDING

dark brown head

speckled upperparts

dark brown back

dark tail

spotted chin, throat, and chest

🟤 BREEDING

mottled underparts

dark face patch

white collar

🟤 NONBREEDING

VOICE High-pitched, squealing, slightly descending *kleeer* in flight.
NESTING In northern part of its range, on island mountainsides; in the south, on tree limbs in old-growth forests; 1 egg; 1 brood; Apr–Sep.
FEEDING Small fish and crustaceans in shallow offshore waters; feeds at night, in pairs.
HABITAT Coniferous forests near coasts while breeding; feeds in shallow coastal waters.
LENGTH 10in (26cm)
WINGSPAN 16in (41cm)

Ancient Murrelet

Synthliboramphus antiquus

Of the five murrelets that occur regularly in North America, this little species is the most numerous. The Ancient Murrelet usually raises two chicks, and takes them out to sea when they are a few days old, usually under the cover of darkness. It can also leap straight out of the sea and into flight. White eyebrow-like plumes on the head give the bird its supposedly "ancient" appearance.

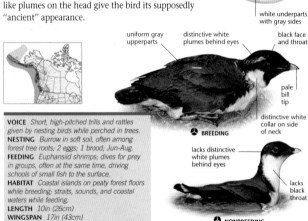

🟤 BREEDING

white under wing

white underparts with gray sides

uniform gray upperparts

distinctive white plumes behind eyes

black face and throat

pale bill tip

distinctive white collar on side of neck

🟤 BREEDING

lacks distinctive white plumes behind eyes

lacks black throat

🟤 NONBREEDING

VOICE Short, high-pitched trills and rattles given by nesting birds while perched in trees.
NESTING Burrow in soft soil, often among forest tree roots; 2 eggs; 1 brood; Jun-Aug.
FEEDING Euphansiid shrimps; dives for prey in groups, often at the same time, driving schools of small fish to the surface.
HABITAT Coastal islands on peaty forest floors while breeding; straits, sounds, and coastal waters while feeding.
LENGTH 10in (26cm)
WINGSPAN 17in (43cm)

Cassin's Auklet ⓢ

Ptychoramphus aleuticus

This seabird usually nests in an underground burrow, which can take a breeding pair many weeks to scratch out. Parent birds fish by day, returning to the nest in the safety of darkness to avoid predators. Nestlings encourage regurgitation by nibbling at a white spot at the base of the parent's lower mandible. Cassin's Auklet has been known to raise more than one brood in a season.

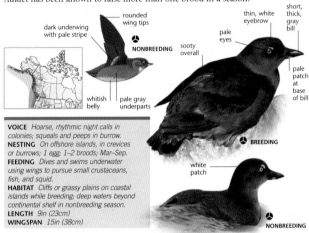

rounded wing tips

dark underwing with pale stripe

🐾 NONBREEDING

whitish belly

pale gray underparts

thin, white eyebrow

pale eyes

sooty overall

short, thick, gray bill

pale patch at base of bill

🐾 BREEDING

white patch

🐾 NONBREEDING

VOICE *Hoarse, rhythmic night calls in colonies; squeals and peeps in burrow.*
NESTING *On offshore islands, in crevices or burrows; 1 egg; 1–2 broods; Mar–Sep.*
FEEDING *Dives and swims underwater using wings to pursue small crustaceans, fish, and squid.*
HABITAT *Cliffs or grassy plains on coastal islands while breeding; deep waters beyond continental shelf in nonbreeding season.*
LENGTH *9in (23cm)*
WINGSPAN *15in (38cm)*

Rhinoceros Auklet ⓢ

Cerorhinca monocerata

Closely related to puffins, the Rhinoceros Auklet is the only auk with a prominent "horn" on top of its bill. It forages closer to shore than puffins, and usually returns to its nesting colonies at night. This trusting seabird often allows boats to approach very closely. When fishing, it carries its catch in its beak, rather than in a throat pouch like other auks.

dark wings

🐾 NONBREEDING

dark upperparts

thin, white plumes curving back

horny structure at base of upper bill

white belly

🐾 BREEDING

lacks facial plumes

smaller bill

🐾 NONBREEDING

VOICE *Series of low, mooing calls, as well as short barks and groans.*
NESTING *Cup of moss or twigs on islands, under vegetation, in crevice or long burrow; 1 egg; 1 brood; Apr–Sep.*
FEEDING *Forages underwater during breeding season, looks for small schooling fish for nestlings; also crustaceans.*
HABITAT *Temperate North Pacific waters, generally far out to sea; nests on islands.*
LENGTH *15in (38cm)*
WINGSPAN *22in (56cm)*

Atlantic Puffin

S

Fratercula arctica

With its black-and-white "tuxedo," ungainly upright posture, and enormous, colorful bill, the Atlantic Puffin is often known as the "clown of the sea." It is seen in summer, when large breeding colonies gather on remote, rocky islands. To feed itself and its young, it can dive down to 200ft (60m) with partly folded wings in pursuit of small schooling fish. One bird was seen holding 83 sand eels in its beak at once. The Atlantic Puffin is the provincial bird of Newfoundland and currently still for Labrador.

short
tail

✦ BREEDING

blue-gray,
orange, and
red stripes
on bill

dull
bill

dusky
gray
face

black back,
collar, and
underwings

✦ NONBREEDING

orange legs
and feet

✦ BREEDING

red
eye-ring

large, colorful,
triangular bill

gray
face

thick
black
line

stocky, rounded
body

white
breast

✦ BREEDING

VOICE *Rising and falling buzzy growl, resembling a chainsaw.*
NESTING *Underground burrow or deep rock crevice lined
with grass and feathers; 1 egg; 1 brood; Jun–Aug.*
FEEDING *Dives for capelin, herring, hake, sand lance, and
other small fish, swallowing underwater, or storing crossways in its
bill to take back to its young.*
HABITAT *Small, rocky, offshore islands while breeding; high seas
far offshore in nonbreeding season.*
LENGTH *12½in (32cm)*
WINGSPAN *21in (53cm)*

Horned Puffin   Ⓢ

Fratercula corniculata

Similar to the Atlantic Puffin in appearance and behavior, the Horned Puffin is larger and lives in the northern Pacific and Bering Sea, nesting on even more remote rocky offshore islands than its Atlantic relative. Outside the breeding season, Horned Puffins stay far out at sea. Upon returning to their breeding grounds, pairs often head straight for the same rock crevice they nested in the year before.

dark wing

gray face

no fleshy "horn" above eye

brown base to bill

Ⓝ **NONBREEDING**

white face

black neck collar

fleshy "horn" above eye

large, yellow bill, with orange tip

white underparts

bright orange legs and toes

dark upperparts

Ⓑ **BREEDING**

VOICE *Low-pitched, rumbling growls in rhythmic phrases.*
NESTING *Deep rock crevices lined with grass and feathers; 1 egg; 1 brood; May–Aug.*
FEEDING *Dives for herring, sand lance, capelin, smelt, and other small fishes to feed to young; squid, crustaceans, and marine worms, eaten underwater.*
HABITAT *Rocky offshore islands while breeding and feeding; winters far out to sea.*
LENGTH *15in (38cm)*
WINGSPAN *23in (59cm)*

Tufted Puffin   Ⓢ

Fratercula cirrhata

Tufted Puffins may be spotted hopping over rocky ledges, sitting alone on the sea, paddling along the surface before taking off, or flying only a couple of feet above the water. This bird's name arises from the curly golden plumes of feathers that adorn its head during the breeding season. It is the largest of the three puffin species.

lacks long golden head plumes

dark face

Ⓝ **NONBREEDING**

no plumes

yellow bill

Ⓟ **POSTBREEDING**

long golden plumes on back of head and nape

large, rounded head

stocky black body

rounded wings

white face

orange bill

dark underparts

orange legs and feet

Ⓑ **BREEDING**

VOICE *Low, moaning growl given from burrow.*
NESTING *Chamber, lined with grass or feathers, at end of tunnel, under rocks, or in burrow; 1 egg; 1 brood; May–Aug.*
FEEDING *Dives for small fish (sand lance, juvenile pollock, capelin); prey consumed underwater or taken ashore to feed young.*
HABITAT *Rocky, treeless islands and coastal cliffs while breeding; winters at sea over deep waters.*
LENGTH *15in (38cm)*
WINGSPAN *25in (64cm)*

Pigeons & Doves

Pigeons and doves are all fairly heavy, plump birds with relatively small heads and short necks. They also possess slender bills with their nostrils positioned in a bumpy mound at the base. Members of this family are powerful and agile fliers. When alarmed, they burst into flight with their wings emitting a distinctive clapping or swishing sound. Pigeons and doves produce a nutritious "crop-milk," which they secrete to feed their young. Despite human activity having severely affected members of this family in the past, the introduced Rock Pigeon has adapted and proliferated worldwide, as has the recently introduced Eurasian Collared-Dove. Among the species native to North America, only the elegant Mourning Dove is as widespread as the various species of introduced birds.

DOVE IN THE SUN
The Mourning Dove sunbathes each side of its body in turn, its wings and tail outspread.

Rock Pigeon Ⓢ

Columba livia

The Rock Pigeon was introduced to the Atlantic coast of North America by 17th century colonists. Now feral, this species is found all over the continent, especially around farms, cities, and towns. It comes in a wide variety of plumage colors and patterns, including bluish gray, checkered, rusty red, and nearly all-white. Its wings usually have two dark bars on them—unique among North American pigeons.

black wing bars

white underwings

iridescence on neck

gray back

white rump

two black wing bars

short bill

dark-tipped tail

Ⓐ ANCESTRAL FORM

variably colored body

no wing bars

Ⓐ FERAL

VOICE *Soft, gurgling coo, roo-c'too-coo, for courtship and threat.*
NESTING *Twig nest on flat, sheltered surface, such as caves, rocky outcrops, and buildings; 2 eggs; several broods; year-round.*
FEEDING *Seeds, fruit, and rarely insects; human foods such as popcorn, bread, peanuts; various farm crops in rural areas.*
HABITAT *Nests in human structures of all sorts; urban areas, farmland, and rocky cliffs.*
LENGTH 11–14in (28–36cm)
WINGSPAN 20–26in (51–67cm)

Band-tailed Pigeon Ⓓ

Patagioenas fasciata

The Band-tailed Pigeon is similar to the Rock Pigeon in its size, posture, body movements, and breeding and feeding behavior. However, in Canada the Band-tailed Pigeon's distribution is limited to the wet coastal forests of the West Coast. It is distinguished by a yellow bill and legs, a white band just above the iridescent green patch on the back of its neck, and a banded tail.

wide tail band

dark gray outer wings

white band on nape

light gray inner wings

dark-tipped yellow bill

iridescence on hind neck

tip of tail pale

uniform blue-gray underparts

yellow legs and toes

gray tail

VOICE *Often silent; series of two-noted, low-frequency whooos punctuated with a pause.*
NESTING *Flat, flimsy platform of twigs, needles, and moss in trees; 1 egg; 1 brood; Apr–Oct.*
FEEDING *Grain, seeds, fruit, acorns, and pine nuts; hangs upside down by its toes from branches to eat dangling nuts and flowers.*
HABITAT *Temperate conifer rainforest, and mountain conifer and mixed-species forests; urban and rural areas.*
LENGTH 13–16in (33–41cm)
WINGSPAN 26in (66cm)

Eurasian Collared-Dove (S)

Streptopelia decaocto

The Eurasian Collared-Dove is easily recognized by the black collar on the back of its neck and its square tail. Introduced in the Bahamas in the mid-1970s, this species is spreading rapidly across the continental mainland. It regularly nests and feeds in urban areas. Based on sightings from locations all over North America, the Eurasian Collared-Dove will likely be regarded as a common species.

black collar on hind neck

pale gray body

dark bill

square tail

gray undertail wing feathers

gray wing feathers

dark outer wing feathers

VOICE *Repeated four-note coo-hoo-HOO-cook, low-pitched; harsh, nasal krreeew in flight.*
NESTING *Platform of twigs, stems, and grasses in trees or on buildings; 2 eggs; multiple broods; Mar–Nov.*
FEEDING *Seed and grain, plant stems and leaves, berries, and some invertebrates.*
HABITAT *Suburban and urban areas (though not large cities); agricultural areas with deciduous trees.*
LENGTH 11½–12in (29–30cm)
WINGSPAN 14in (35cm)

Mourning Dove (S)

Zenaida macroura

One of the most familiar of North American birds, the Mourning Dove has a grayish tan body with a pale, rosy breast and black spots on folded wings, and is well known to those who live on farms and in suburbia. There are two subspecies—the larger grayish brown *Z. m. carolinensis* in the east, and the smaller, paler *Z. m. marginella* in the west.

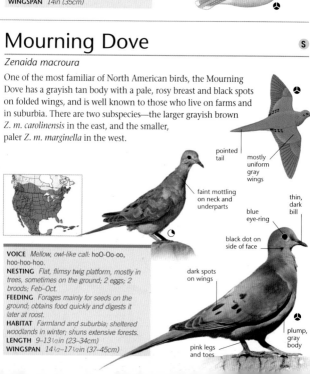

pointed tail

mostly uniform gray wings

faint mottling on neck and underparts

thin, dark bill

blue eye-ring

black dot on side of face

dark spots on wings

black spots on wings

pink legs and toes

plump, gray body

VOICE *Mellow, owl-like call: hoO-Oo-oo, hoo-hoo-hoo.*
NESTING *Flat, flimsy twig platform, mostly in trees, sometimes on the ground; 2 eggs; 2 broods; Feb–Oct.*
FEEDING *Forages mainly for seeds on the ground; obtains food quickly and digests it later at roost.*
HABITAT *Farmland and suburbia; sheltered woodlands in winter; shuns deciduous forests.*
LENGTH 9–13½in (23–34cm)
WINGSPAN 14½–17½in (37–45cm)

Cuckoos

Cuckoos are notorious for laying eggs in other birds' nests, but the two species found in Canada seldom do so. Generally shy and reclusive, the Black-billed Cuckoo and Yellow-billed Cuckoo favor dense, forested habitats. Both species usually build a nest and raise their own offspring, but occasionally these two cuckoo species lay their eggs in other birds' nest, including each other's, and even in the nests of their own kind. In flight, cuckoos are often mistaken for small birds of prey. While they sometimes pounce on lizards, frogs, and other small animals, they mostly glean insects from the foliage of trees. Besides their slender bodies and long tails, cuckoos have zygodactyl feet, with the two inner toes pointing forward and the two outer toes pointing backward.

NIGHT SINGER
During the breeding season, the Black-billed Cuckoo will often call throughout the night.

Black-billed Cuckoo

(S)

Coccyzus erythropthalmus

The Black-billed Cuckoo is usually difficult to spot because of its secretive nature and dense, leafy habitat. This species feeds mainly on spiny caterpillars, but the spines of these insects can become lodged in the cuckoo's stomach, obstructing digestion, so the bird periodically sloughs off its stomach lining to clear it. During the breeding season, the birds call throughout the night.

small white spots on tips of tail feathers

long tail

long wings

bare red skin around eye

long, black, decurved bill

grayish brown back

pale grayish white underparts

grayish feet

long tail

VOICE Series of 2–5 repeatedly whistled notes, coo-coo-coo-coo, with short breaks between series.
NESTING Shallow cup of sticks lined with moss, leaves, grass, and feathers; 2–4 eggs; 1 brood; May–Jul.
FEEDING Almost exclusively eats caterpillars, especially tent caterpillars and gypsy moths.
HABITAT Thickly wooded areas near water; brushy forest edges and evergreen woods.
LENGTH 11–12in (28–31cm)
WINGSPAN 16–19in (41–48cm)

Yellow-billed Cuckoo

(S)

Coccyzus americanus

The Yellow-billed Cuckoo is a shy, slow-moving bird with a habit of calling more often on cloudy days, earning it the nickname "rain crow." In addition to raising young in its own nest, females often lay eggs in the nests of other species, including the Black-billed Cuckoo. The host species may be chosen on the basis of how closely the color of its eggs matches those of its own.

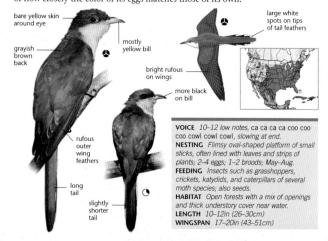

bare yellow skin around eye

grayish brown back

mostly yellow bill

large white spots on tips of tail feathers

bright rufous on wings

more black on bill

rufous outer wing feathers

long tail

slightly shorter tail

VOICE 10–12 low notes, ca ca ca ca coo coo coo cowl cowl cowl, slowing at end.
NESTING Flimsy oval-shaped platform of small sticks, often lined with leaves and strips of plants; 2–4 eggs; 1–2 broods; May–Aug.
FEEDING Insects such as grasshoppers, crickets, katydids, and caterpillars of several moth species; also seeds.
HABITAT Open forests with a mix of openings and thick understory cover near water.
LENGTH 10–12in (26–30cm)
WINGSPAN 17–20in (43–51cm)

Owls

Most owls are active primarily at night and have developed adaptations for living in low-light environments. Their large eyes are sensitive enough to see in the dark and face forward to maximize binocular vision. Since the eyes are somewhat fixed in their sockets, a flexible neck allows the owls to turn their heads almost 270 degrees. Ears are offset on each side of the head to help identify the source of a sound. Many owls have serrations on the forward edges of their flight feathers to cushion airflow, so their flight is silent while stalking prey. All North American owls are predatory to some degree and they inhabit most areas of the continent. The Burrowing Owl is unique in that it hunts during the day and nests underground.

KEEN HEARING
The Great Gray Owl can hunt by sound alone, allowing it to locate and capture prey hidden even beneath a thick snow cover.

Barn Owl

Tyto alba

Aptly named, the Barn Owl inhabits old sheds, sheltered rafters, and empty buildings in rural fields. It is secretive and primarily nocturnal, flying undetected until its screeching call pierces the air. The Eastern populations are endangered due to modern farming practices, which have decimated prey populations and reduced the number of barns for nesting.

barring on wings and tail

head lacks "ear" tufts

long wings

rounded, heart-shaped facial disc

ruff surrounds facial disk

pale buff upperparts

gray and black spots

relatively small, dark eyes

white underparts

feathered legs

VOICE *Loud, raspy, screeching shriek, shkreee, often in flight; clicking sounds associated with courtship.*
NESTING *Unlined cavity in tree, cave, building, hay bale, or nest box; 5–7 eggs; 1–2 broods; Mar–Sep.*
FEEDING *Hunts on the wing for small rodents such as mice.*
HABITAT *Open habitats such as desert, grassland, and fields.*
LENGTH *12½–15½in (32–40cm)*
WINGSPAN *3ft 3in (100cm)*

Flammulated Owl

Otus flammeolus

The tiny, nocturnal Flammulated Owl nests in dry mountain pine forests from British Columbia to Mexico, moving south to Central America for the winter months. Its dark, watery-looking eyes distinguish it from other species of small North American owls. This species appears to breed in loose colonies, although this may reflect patchiness in habitat quality.

short tail

long, rounded wings

tawny underwings

tawny "shoulder" bar

smaller in size than gray form

RED FORM

reddish brown facial disc

small "ear" tufts, often hidden

dark eyes

grayish brown body

dark streaks on underparts

GRAY FORM

VOICE *Soft low-frequency toots, often difficult to locate; barks and screams when disturbed.*
NESTING *Cavity in tree, woodpecker hole, nest box; 3–4 eggs; 1–2 broods; May–Aug.*
FEEDING *Flies from stationary perch to capture insects—mostly moths, and beetles— from branches, foliage, or ground.*
HABITAT *Semi-arid mountain forests with Ponderosa and Yellow Pine; open wooded areas with clearings.*
LENGTH *6–6¾in (15–17cm)*
WINGSPAN *16in (41cm)*

Eastern Screech-Owl ⓢ

Megascops asio

This widespread little owl has adapted to suburban areas, and its distinctive call is a familiar sound across southern Canada at almost any time of the year. An entirely nocturnal species, it may be found roosting during the day in a birdhouse or tree cavity. With gray and red color forms, this species shows more plumage variation than the Western Screech-Owl.

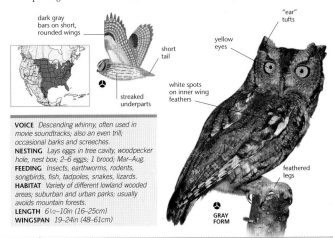

dark gray bars on short, rounded wings

short tail

streaked underparts

"ear" tufts

yellow eyes

white spots on inner wing feathers

feathered legs

GRAY FORM

VOICE *Descending whinny, often used in movie soundtracks; also an even trill; occasional barks and screeches.*
NESTING *Lays eggs in tree cavity, woodpecker hole, nest box; 2–6 eggs; 1 brood; Mar–Aug.*
FEEDING *Insects, earthworms, rodents, songbirds, fish, tadpoles, snakes, lizards.*
HABITAT *Variety of different lowland wooded areas; suburban and urban parks; usually avoids mountain forests.*
LENGTH *6½–10in (16–25cm)*
WINGSPAN *19–24in (48–61cm)*

Western Screech-Owl ⓣ

Megascops kennicottii

The Western Screech-Owl lives in a wide variety of wooded areas, including suburban habitats. Because of its nocturnal habits, the Western Screech-Owl is heard more often than it is seen; its "bouncing ball" call, sometimes repeated for hours, is a familiar sound in much of western North America. This species exhibits significant differences in plumage color, and size, depending on its geographical location.

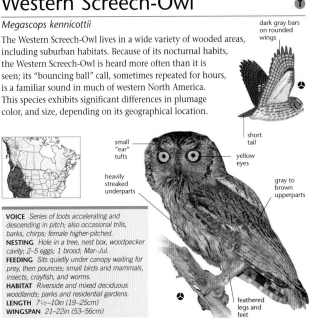

dark gray bars on rounded wings

short tail

small "ear" tufts

heavily streaked underparts

yellow eyes

gray to brown upperparts

feathered legs and feet

VOICE *Series of toots accelerating and descending in pitch; also occasional trills, barks, chirps; female higher-pitched.*
NESTING *Hole in a tree, nest box, woodpecker cavity; 2–5 eggs; 1 brood; Mar–Jul.*
FEEDING *Sits quietly under canopy waiting for prey, then pounces; small birds and mammals, insects, crayfish, and worms.*
HABITAT *Riverside and mixed deciduous woodlands; parks and residential gardens.*
LENGTH *7½–10in (19–25cm)*
WINGSPAN *21–22in (53–56cm)*

Snowy Owl

(S)

Bubo scandiacus

An icon of the far north, the Snowy Owl is a bird of the open tundra, where it hunts from headlands or hummocks and nests on the ground. In such a harsh environment, the Snowy Owl largely depends on lemmings for prey. It is fiercely territorial, and will valiantly defend its young in the nest even against larger animals, such as the Arctic Fox. The Snowy Owl is the provincial bird of Quebec.

white face

flecked gray-brown

dusky barring

variably barred underparts

large round head

yellow eyes

variable barring on wings

nearly all-white breast

feathered legs

VOICE *Deep hoots, doubled or given in a short series, usually by male; also rattles, whistles, and hisses.*
NESTING *Scrape in ground vegetation or dirt, with no lining; 3–12 eggs; 1 brood; May–Sep.*
FEEDING *Mostly hunts lemmings, but takes whatever small mammals and birds it can find, and even occasionally fish.*
HABITAT *Tundra while breeding; open, tree-less spaces such as dunes, marshes, and airfields.*
LENGTH *20–27in (51–68cm)*
WINGSPAN *4¼–5¼aft (1.3–1.6m)*

Great Horned Owl

S

Bubo virginianus

The Great Horned Owl is perhaps the archetypal owl. Large and adaptable, it is resident from Alaska to Tierra del Fuego and is the provincial bird of Alberta. With such a big range, geographical variation occurs; at least 13 subspecies have been described. The Great Horned Owl's deep hoots are easily recognized, and can often be heard in movie soundtracks. The bird is the top predator in its food chain, often killing and eating other owls, and even skunks. An early breeder and a fierce defender of its young, it starts hooting in the middle of winter, and often lays its eggs in January.

large "ears"

yellow eye

rusty facial disk

long, broad wing

dark arc on wing

heavy barring of underparts

white throat and chin

mottled, barred, brownish and gray upperparts

barring on undertail

barred underparts

VOICE *Series of hoots whoo-hoo-oo-o; also screams, barks, and hisses; female higher-pitched.*
NESTING *Old stick nest, in tree, exposed cavity, cliff, human structure, or on the ground; 1–5 eggs; 1 brood; Jan–Apr.*
FEEDING *Mammals, reptiles, amphibians, birds, and insects; mostly nocturnal.*
HABITAT *Prefers fragmented landscapes: desert, swamp, prairie, woodland, and urban areas.*
LENGTH *18–25in (46–63cm)*
WINGSPAN *3–5ft (0.9–1.6m)*

Spotted Owl

Strix occidentalis

The Spotted Owl is seriously threatened by competition from expanding Barred Owl populations and habitat loss by clearcutting of old-growth forests. It has been divided into four subspecies: *S. o. caurina* can be found in British Columbia, the Pacific Northwest, and northern California. Spotted and Barred Owls also interbreed, producing hybrids.

pale underwings

large, puffy head, without ear tufts

dark eye

pale bill

rich, dark brown upperparts with white spots

NORTHWEST US, CANADA

pale oval bars on underparts

fluffy down feathers around head

VOICE *Typical call* whoo hoo-hoo hooo, *with emphasis on last syllable; whistles and barks.*
NESTING *No obvious nest; lays eggs in broken-off snags, cavities, and platforms, occasionally cliffs; 1–4 eggs; 1 brood; Mar–Jul.*
FEEDING *Sits, waits and pounces on prey; eats small rodents.*
HABITAT *Old-growth and mature stands of fir, hemlock, redwood, pine, cedar, oak, and mixed riverside woodlands.*
LENGTH *18–19in (46–48cm)*
WINGSPAN *3½ft (1.1m)*

Great Gray Owl

Strix nebulosa

The Great Gray Owl has a thick layer of feathers that insulate it against cold northern winters. Often able to detect prey by sound alone, it will plunge through deep snow, or into a burrow, to snatch unseen prey. This somewhat nomadic bird, prone to occasional invasions in the south, may also hunt by daylight, usually at dawn or dusk. The Great Gray Owl is the provincial bird of Manitoba.

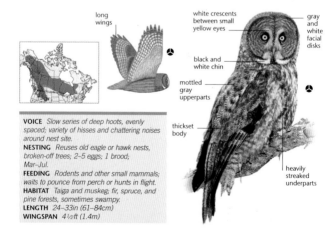

long wings

white crescents between small yellow eyes

gray and white facial disks

black and white chin

mottled gray upperparts

thickset body

heavily streaked underparts

VOICE *Slow series of deep hoots, evenly spaced; variety of hisses and chattering noises around nest site.*
NESTING *Reuses old eagle or hawk nests, broken-off trees; 2–5 eggs; 1 brood; Mar–Jul.*
FEEDING *Rodents and other small mammals; waits to pounce from perch or hunts in flight.*
HABITAT *Taiga and muskeg; fir, spruce, and pine forests, sometimes swampy.*
LENGTH *24–33in (61–84cm)*
WINGSPAN *4½ft (1.4m)*

Barred Owl

S

Strix varia

The Barred Owl is more adaptable and aggressive than its close relative, the Spotted Owl. The former's recent range expansion has brought the two species into closer contact, which has led to Barred Owls displacing and even killing Spotted Owls, as well as occasional interbreeding. The Barred Owl is mostly nocturnal, but may also call or hunt during the day.

rounded wings

large, round head

brown upperparts

heavy white spotting

dark eyes

conspicuously yellowish bill

barred tail

barring on breast

streaking on belly

VOICE *Series of hoots in rhythm: who-cooks-for-you, who-cooks-for-you-all; also pair duetting, cawing, cackling, and guttural sounds.*
NESTING *No obvious nest; lays eggs in broken-off branches, cavities, old stick nests; 1–5 eggs; 1 brood; Jan–Sep.*
FEEDING *Perches then pounces; small mammals, birds, amphibians, reptiles, insects, and spiders.*
HABITAT *Variety of wooded habitats such as conifer rainforest, mixed hardwoods, and cypress swamps.*
LENGTH *17–19½in (43–50cm)*
WINGSPAN *3½ft (1.1m)*

Northern Hawk Owl (S)

Surnia ulula

The Northern Hawk Owl is streamlined, a powerful flier, and an active daytime hunter. It is patchily distributed across northern Canadian forests, far from most human settlements, and thus, is seldom seen on its breeding grounds. In winter, the bird is somewhat nomadic, and is occasionally seen south of its breeding range for a few days or weeks in southern Canada and the northern US.

long tail

long wings

patterned face

fine spotting on forehead and crown

yellowish eyes

black line around white face

brownish black upperparts

heavy white marking

regularly barred underparts

VOICE *Ascending, whistled, drawn-out trill; also chirps, screeches, and yelps.*
NESTING *Cavities, hollows, broken-off branches, old stick nests, nest boxes; 3–13 eggs; 1 brood; Apr–Aug.*
FEEDING *Swoops from an elevated perch to pounce on prey; rodents in summer, and grouse, ptarmigan, and other birds in winter.*
HABITAT *Sparse woodland or mixed conifer forest with swamps, burnt areas, storm damage.*
LENGTH *14–17½in (36–44cm)*
WINGSPAN *31in (80cm)*

Northern Pygmy-Owl (S)

Glaucidium gnoma

The tiny Northern Pygmy-Owl is a fierce hunter, regularly preying on other birds, including relatively large ones such as bobwhites. It is often active during the day, most frequently around dawn and dusk, and in winter can be seen in gardens, pouncing on birds at feeders. It has "false eyes"—a pair of black-feathered spots on the back of its head, which may act as a deterrent to potential attackers.

long tail

rounded wings

brown to gray upperparts with white spots

round head

spotted crown

yellow eye

yellowish bill

heavily streaked whitish underparts

long tail, with brown and white barring

VOICE *Hollow poot, poot, poot, 1–2 seconds apart, continuing in series for minutes or more; also excited trill.*
NESTING *Usually unlined cavity in tree; 2–7 eggs; 1 brood; Apr–Jul.*
FEEDING *Pounces from perch, pinning prey to the ground; insects, reptiles, birds, and mammals.*
HABITAT *Mixed spruce, fir, pine, hemlock, cedar, and oak woodlands in mountains.*
LENGTH *6½–7in (16–18cm)*
WINGSPAN *15in (38cm)*

Burrowing Owl

Athene cunicularia

The Burrowing Owl is unique among North American owls in nesting underground. It mostly uses the abandoned burrows of prairie dogs, ground squirrels, armadillos, badgers, and other mammals, but will excavate its own burrow with its bill and feet. It usually nests in loose colonies, too. Active by day or night, the Burrowing Owl hunts prey on foot or on the wing.

short, rounded wings

short tail

chest spotted with white

yellow eyes

white streaking

white contrasting with dark brown band below

white spots

brown upper-parts with white spotting

brown streaks on lower belly

long, feathered legs

short tail

VOICE Coo-cooo, or ha-haaa; also clucks, chatters, warbles, and screams.
NESTING Cavity lined with grass, feathers, sometimes animal dung, at end of burrow; 8–10 eggs; 1 brood; Mar–Aug.
FEEDING Insects, and occasionally small mammals, birds, reptiles, and amphibians.
HABITAT Wide range of open habitats not prone to flooding: pastures, plains, deserts, grasslands, and steppes.
LENGTH 7½–10in (19–25cm)
WINGSPAN 21½in (55cm)

Boreal Owl

Aegolius funereus

The female Boreal Owl is much bigger than the male. Males will mate with two or three females in years when voles and other small rodents are abundant. It is rarely seen, breeding at high elevations in isolated mountain ranges and hunting at night. White spotting on the crown, a grayish bill, and a black facial disk distinguish the Boreal Owl from the Northern Saw-whet Owl.

rounded wings

white and brown streaked underparts

usually flat-topped head, with fine white spots

yellow eyes

pale bill

black border around face

short tail

VOICE Series of whistles, usually increasing in volume and intensity; screeches and hisses.
NESTING Tree cavity, often woodpecker hole; nest boxes; 3–6 eggs; 1 brood; Mar–Jul.
FEEDING Small mammals; birds and insects; pounces from elevated perch; sometimes stores prey and uses its body to thaw it.
HABITAT Boreal forests of spruce, poplar, aspen, birch, and balsam fir; western populations in subalpine forests of fir, spruce.
LENGTH 8½–11in (21–28cm)
WINGSPAN 21½–24in (54–62cm)

Northern Saw-whet Owl

Aegolius acadicus

One of the most secretive yet common and widespread owls in North America, the Northern Saw-whet Owl is much more often heard than seen. Strictly nocturnal, it is concealed as it sleeps by day in thick vegetation, usually in conifers such as cedars. Although the same site may be used for months if undisturbed, it is not an easy bird to locate. When it is discovered, it "freezes," and relies on its camouflage rather than flying off. At night it watches intently from a perch, before swooping down to snatch its prey.

thin white streaks on forehead and crown

yellow eyes

dark bill

brown streaks

rounded wings

short tail

white patch between eyes

whitish eyebrows

chestnut-brown upperparts with white spots

unmarked white undertail feathers

VOICE *Series of rapid whistled notes, on constant pitch; also whines and squeaks.*
NESTING *Unlined cavity in tree, usually old woodpecker hole or nest box; 4–7 eggs; 1 brood; Mar–Jul.*
FEEDING *Hunts from elevated perch; small mammals, including mice and voles; also insects and small birds.*
HABITAT *Coniferous, mixed deciduous, and swampy forests, wooded wetlands, bogs; open woodlands, shrubby areas in winter.*
LENGTH 7–8½in (18–21cm)
WINGSPAN 16½–19in (42–48cm)

Long-eared Owl

Asio otus

The widespread Long-eared Owl is seldom seen, being secretive and nocturnal. By day it roosts, sometimes colonially, high up and out of sight in thick cover, flying out at nightfall to hunt small mammals on the wing over open areas. Its wing feathers have sound-suppressing structures that allow it to fly almost silently to hear the slightest rustle on the ground below and not alert its prey.

tan patch on outer wing

dark wrist patch

gray tips

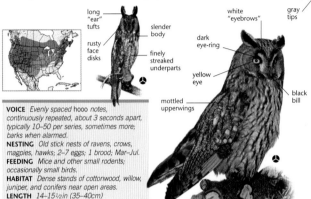

long "ear" tufts

slender body

rusty face disks

finely streaked underparts

white "eyebrows"

dark eye-ring

yellow eye

black bill

mottled upperwings

VOICE *Evenly spaced* hooo *notes, continuously repeated, about 3 seconds apart, typically 10–50 per series, sometimes more; barks when alarmed.*
NESTING *Old stick nests of ravens, crows, magpies, hawks; 2–7 eggs; 1 brood; Mar–Jul.*
FEEDING *Mice and other small rodents; occasionally small birds.*
HABITAT *Dense stands of cottonwood, willow, juniper, and conifers near open areas.*
LENGTH *14–15½in (35–40cm)*
WINGSPAN *34–39in (86–98cm)*

Short-eared Owl

Asio flammeus

This owl is often seen on cloudy days or toward dusk, soaring back and forth low over open fields, sometimes with Northern Harriers. Although territorial in the breeding season, it sometimes winters in communal roosts of up to 200 birds, occasionally alongside Long-eared Owls. Unlike other North American owls, the Short-eared Owl builds its own nest.

black wing tips

row of pale spots along sides of back

dark wrist patch

orange-buff to yellowish outer wings

narrow, dark bar

whitish underwing

white belly

short ear tufts

blackish eye-ring

yellow eyes

pale face disks

large, round head

complex, buff marbling on upperparts

fine dark streaks

whitish buff underparts

VOICE *Usually silent; males: rapid* hoo hoo hoo, *often given during display flights; also barking,* chee-oww.
NESTING *Scrape lined with grass and feathers on ground; 4–7 eggs; 1–2 broods; Mar–Jun.*
FEEDING *Eats small mammals and some birds.*
HABITAT *Open areas: prairie, grasslands, tundra, fields, and marshes.*
LENGTH *13½–16in (34–41cm)*
WINGSPAN *2¾ –3½ft (0.9–1.1m)*

Nightjars

Nightjars are heard more often than they are seen—except for the Common Nighthawk, which regularly forages for insects at dawn and dusk. Nightjars are medium-sized birds that use their long wings and wide tails to make rapid and graceful turns to capture their insect prey in the air. Their wide, gaping mouths are surrounded by bristles that greatly aid in foraging efforts. These birds are well camouflaged with a mottled mixture of various browns, grays, and blacks—a useful trait during the nesting season when all nightjars lay their patterned eggs directly on the ground, without any nest material. The most reliable means of telling species apart is their voice. Most members of the family migrate and move southward as insects become dormant in the North.

SITTING PRETTY
Unusual for birds, members of the nightjar family, such as this Common Nighthawk, often perch lengthwise on branches.

Common Nighthawk

Chordeiles minor

Common Nighthawks are easy to spot as they swoop over parking lots, city streets, and athletics fields during the warm summer months. They are more active at dawn and dusk than at night, pursuing insect prey up to 250ft (76m) in the air. The species once took the name Booming Nighthawk, a reference to the remarkable flight display of the male birds, during which they dive rapidly towards the ground, causing their feathers to vibrate and produce a characteristic "booming" sound.

narrow wings

white wing patch

white bars on outer wing feathers

pointed wings

♂

♂

white throat

very small bill

large, dark eye

long wings

delicate, gray-black pattern overall

♀

barring on gray underparts

VOICE Nasal *peeent*; also soft clucking noises from both sexes.
NESTING Nests on ground on rocks, wood, leaves, or sand, also on gravel-covered rooftops in urban areas; 2 eggs; 1 brood; May–Jul.
FEEDING Catches airborne insects, especially moths, mayflies, and beetles, also ants; predominantly active at dusk and dawn.
HABITAT Wide variety of open habitats such as cleared forests, fields, grassland, beaches, sand dunes; also common in urban areas, including cities.
LENGTH 9–10in (23–26cm)
WINGSPAN 22–24in (56–61cm)

Common Poorwill

S

Phalaenoptilus nuttallii

This nocturnal bird has much shorter wings than its relatives, and a stubbier tail, but a comparatively large head. It is able to go into a state of torpor, similar to mammalian hibernation, and it may remain in this state for several weeks during cold weather when food is unavailable. Males and females look similar, but the male has whitish corners to its tail, while the female's are more buffy.

short tail

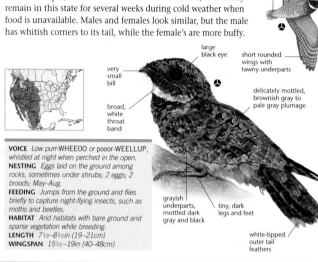

large black eye

very small bill

broad, white throat band

short rounded wings with tawny underparts

delicately mottled, brownish gray to pale gray plumage

grayish underparts, mottled dark gray and black

tiny, dark legs and feet

white-tipped outer tail feathers

VOICE *Low purr-WHEEOO or pooor-WEELLUP, whistled at night when perched in the open.*
NESTING *Eggs laid on the ground among rocks, sometimes under shrubs; 2 eggs; 2 broods; May–Aug.*
FEEDING *Jumps from the ground and flies briefly to capture night-flying insects, such as moths and beetles.*
HABITAT *Arid habitats with bare ground and sparse vegetation while breeding.*
LENGTH *7½–8½in (19–21cm)*
WINGSPAN *15½–19in (40–48cm)*

Whip-poor-will

T

Caprimulgus vociferus

The Whip-poor-will is heard more often than seen. Its camouflage makes it difficult to spot on the forest floor and it usually flies away only when an intruder is very close. While the male feeds the first brood until fledging, the female lays eggs for a second brood. The two eggs from each brood may hatch together during a full moon, allowing the parents more light at night to forage for their young.

rounded wings

♂

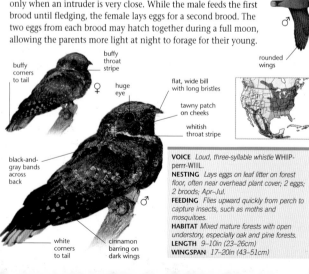

buffy corners to tail

buffy throat stripe

♀

huge eye

flat, wide bill with long bristles

tawny patch on cheeks

whitish throat stripe

black-and-gray bands across back

white corners to tail

cinnamon barring on dark wings

♂

VOICE *Loud, three-syllable whistle WHIP-perrr-WIIL.*
NESTING *Lays eggs on leaf litter on forest floor, often near overhead plant cover; 2 eggs; 2 broods; Apr–Jul.*
FEEDING *Flies upward quickly from perch to capture insects, such as moths and mosquitoes.*
HABITAT *Mixed mature forests with open understory, especially oak and pine forests.*
LENGTH *9–10in (23–26cm)*
WINGSPAN *17–20in (43–51cm)*

Swifts

The most aerial birds in North America—if not the world—swifts eat, drink, court, mate, and even sleep on the wing. Unsurprisingly, swifts also are some of the fastest and most acrobatic flyers of the bird world. Several species have been clocked at over 100mph (160kph). Looking like "flying cigars," they feed on insects caught in zooming, zigzagging and dashing pursuits.

ACROBATIC FLOCKS
White-throated Swifts are usually seen in groups of a handful to hundreds of birds.

Hummingbirds

Hummingbirds are sometimes referred to as the crown jewels of the bird world. Most North American male hummingbirds have a colorful throat patch called a gorget, but most females lack this attribute. Hummingbirds can fly backwards, sideways, up, down, and hover due to their unique figure-eight, rapid wing strokes and reduced wing bone structure. Their long, thin bills allow them access to nectar in tubular flowers.

NECTAR FEEDERS
All North American hummingbirds, such as this Black-chinned, subsist on nectar from wildflowers.

Black Swift

(S)

Cypseloides niger

The largest of the North American swifts, the Black Swift forages at high altitudes and nests on sea cliffs or behind waterfalls in mountainous terrains. On cold and cloudy days, they may be easier to see as they forage lower for their aerial insect prey. They often form large feeding flocks, particularly when swarms of winged ants occur.

pale underwings

black body

slightly notched tail

grayish black head

black eye

blackish upperparts

black patch in front of eye

black tail lacks "spines"

dark plumage with glossy sheen

very long, sickle-shaped wings

VOICE *Generally silent; twittering chips during interactions with other swifts; sharp cheep when approaching nest.*
NESTING *Shallow cup of moss and mud on ledge or in rocky niche, often behind waterfalls; 1 egg; 1 brood; Jun–Sep.*
FEEDING *Airborne flies, beetles, bees, spiders, and other arthropods on the wing.*
HABITAT *Mountains from May–Jun to early Oct, feeding high over habitat near nesting sites.*
LENGTH *7in (18cm)*
WINGSPAN *18in (46cm)*

Vaux's Swift

(S)

Chaetura vauxi

This acrobatic, fast-flying bird is North America's smallest swift. Its western range, small size, rapid and fluttering flight, and distinctive shape help identify this species. Vaux's Swifts are typically found foraging in flocks over mature forest and can be spotted on cold, cloudy days, often mixed with other swifts. Very large flocks are also sighted "pouring" into communal roost sites at dusk.

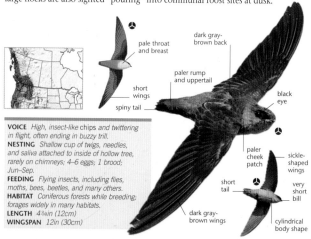

pale throat and breast

short wings

spiny tail

dark gray-brown back

paler rump and uppertail

black eye

paler cheek patch

sickle-shaped wings

short tail

very short bill

dark gray-brown wings

cylindrical body shape

VOICE *High, insect-like chips and twittering in flight, often ending in buzzy trill.*
NESTING *Shallow cup of twigs, needles, and saliva attached to inside of hollow tree, rarely on chimneys; 4–6 eggs; 1 brood; Jun–Sep.*
FEEDING *Flying insects, including flies, moths, bees, beetles, and many others.*
HABITAT *Coniferous forests while breeding; forages widely in many habitats.*
LENGTH *4¾in (12cm)*
WINGSPAN *12in (30cm)*

Chimney Swift

Chaetura pelagica

Nicknamed "spine-tailed," the Chimney Swift is a familiar summer sight and sound, racing through the skies east of the Rockies. These birds do almost everything on the wing—feeding, drinking, and even bathing. Chimney Swifts have adapted to nest in human structures, including real and even provided fake chimneys, although they once nested in tree holes. It has expanded its range west and south.

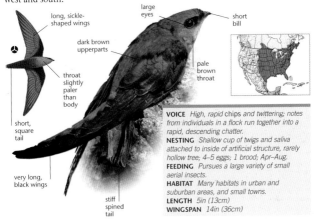

long, sickle-shaped wings

large eyes

short bill

dark brown upperparts

pale brown throat

throat slightly paler than body

short, square tail

very long, black wings

stiff spined tail

VOICE *High, rapid chips and twittering; notes from individuals in a flock run together into a rapid, descending chatter.*
NESTING *Shallow cup of twigs and saliva attached to inside of artificial structure, rarely hollow tree; 4–5 eggs; 1 brood; Apr–Aug.*
FEEDING *Pursues a large variety of small aerial insects.*
HABITAT *Many habitats in urban and suburban areas, and small towns.*
LENGTH *5in (13cm)*
WINGSPAN *14in (36cm)*

White-throated Swift

Aeronautes saxatalis

Often seen racing around the cliffs on which they nest, White-throated Swifts can be identified by their black-and-white plumage and longer tail. This species has become increasingly common in urban areas, as it has adapted to nesting in human structures such as bridges and quarries. Huge flocks of White-throated Swifts can be seen rushing into communal roosts at dusk.

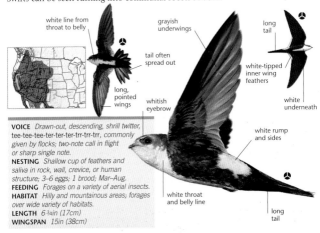

white line from throat to belly

grayish underwings

long tail

tail often spread out

white-tipped inner wing feathers

long, pointed wings

whitish eyebrow

white underneath

white rump and sides

white throat and belly line

long tail

VOICE *Drawn-out, descending, shrill twitter, tee-tee-tee-ter-ter-ter-ter-trr-trr-trr, commonly given by flocks; two-note call in flight or sharp single note.*
NESTING *Shallow cup of feathers and saliva in rock, wall, crevice, or human structure; 3–6 eggs; 1 brood; Mar–Aug.*
FEEDING *Forages on a variety of aerial insects.*
HABITAT *Hilly and mountainous areas; forages over wide variety of habitats.*
LENGTH *6¾in (17cm)*
WINGSPAN *15in (38cm)*

Ruby-throated Hummingbird ⓢ

Archilochus colubris

The Ruby-throated Hummingbird is easily identified in most of its range, though more difficult to distinguish in areas where other species are found, particularly during migration. Before migration, these birds add about 1/16oz (2g) of fat to their weight to provide enough fuel for their nonstop 800-mile (1,300km) flight across the Gulf of Mexico.

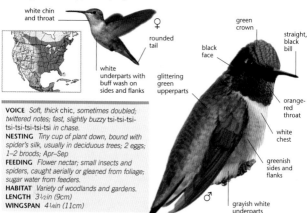

♂
dark, forked tail

white chin and throat
♀
rounded tail

white underparts with buff wash on sides and flanks

glittering green upperparts

green crown
black face
straight, black bill

orange-red throat

white chest

greenish sides and flanks

grayish white underparts
♂

VOICE *Soft, thick* chic, *sometimes doubled; twittered notes; fast, slightly buzzy* tsi-tsi-tsi-tsi-tsi-tsi-tsi-tsi *in chase.*
NESTING *Tiny cup of plant down, bound with spider's silk, usually in deciduous trees; 2 eggs; 1–2 broods; Apr–Sep.*
FEEDING *Flower nectar; small insects and spiders, caught aerially or gleaned from foliage; sugar water from feeders.*
HABITAT *Variety of woodlands and gardens.*
LENGTH *3½in (9cm)*
WINGSPAN *4¼in (11cm)*

Black-chinned Hummingbird ⓢ

Archilochus alexandri

The Black-chinned Hummingbird is found in southern British Columbia and is widespread across the western US. The iridescent purple border on the black chin is only visible in the right light. During courtship, the males perform a distinctive dive display comprising several broad arcs in addition to a short, back-and-forth shuttle display, producing a droning noise by the bird's wings.

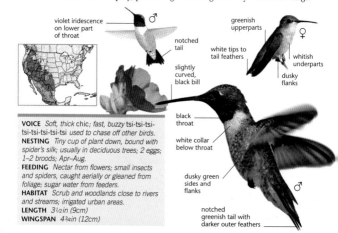

violet iridescence on lower part of throat
♂
notched tail
slightly curved, black bill

greenish upperparts
white tips to tail feathers
♀
whitish underparts
dusky flanks

black throat
white collar below throat

dusky green sides and flanks
♂

notched greenish tail with darker outer feathers

VOICE *Soft, thick* chic; *fast, buzzy* tsi-tsi-tsi-tsi-tsi-tsi-tsi-tsi *used to chase off other birds.*
NESTING *Tiny cup of plant down, bound with spider's silk; usually in deciduous trees; 2 eggs; 1–2 broods; Apr–Aug.*
FEEDING *Nectar from flowers; small insects and spiders, caught aerially or gleaned from foliage; sugar water from feeders.*
HABITAT *Scrub and woodlands close to rivers and streams; irrigated urban areas.*
LENGTH *3½in (9cm)*
WINGSPAN *4¾in (12cm)*

Anna's Hummingbird

Calypte anna

The most common garden hummingbird along the Pacific Coast from British Columbia to Baja California, the iridescent rose-red helmet of a male Anna's Hummingbird is spectacular and distinctive. The females are rather drab by comparison. This adaptable hummingbird has expanded its range dramatically in the last century because of the availability of garden flowers and feeders. It previously bred only in areas of dense evergreen shrubs along the coast of southern California.

green crown and nape

♀

pale throat

♂

reddish spots or flecks on throat

square tail

short, straight, black bill

rose-red head, sides of neck, and throat

pale gray underparts

rounded, green tail

iridescent green upperparts

slightly notched, dark green tail

grayish underparts

greenish sides and flanks

♂

VOICE *Hard, sharp* tsit, *often doubled or in series when perched; fast, buzzy chatter in chase; song variable series of thin, high, buzzing, warbled notes.*
NESTING *Tiny cup of mostly plant down, with lichen on the exterior, bound with spider's silk, built in trees or shrubs; 2 eggs; 2 broods; Dec–Jul.*
FEEDING *Nectar from flowers; small insects and spiders.*
HABITAT *Coastal dense shrubs and open woodland while breeding.*
LENGTH 4in (10cm)
WINGSPAN 5in (13cm)

Rufous Hummingbird (S)

Selasphorus rufus

The Rufous Hummingbird is aggressive despite its small size; it often chases other hummingbirds away from nectar sources. It also breeds farther north than any other North American species of hummingbird and undertakes a lengthy migration. Males are recognizable by their overall fiery orange-rufous color, but females and immature birds are more difficult to distinguish.

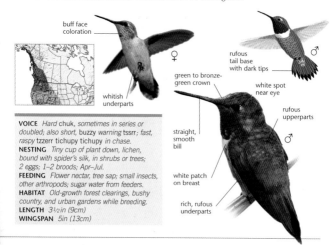

buff face coloration

♀

whitish underparts

rufous tail base with dark tips

♂

green to bronze-green crown

white spot near eye

rufous upperparts

straight, smooth bill

♂

white patch on breast

rich, rufous underparts

VOICE Hard chuk, sometimes in series or doubled; also short, buzzy warning tssrr; fast, raspy tzzerr tichupy tichupy in chase.
NESTING Tiny cup of plant down, lichen, bound with spider's silk, in shrubs or trees; 2 eggs; 1–2 broods; Apr–Jul.
FEEDING Flower nectar, tree sap; small insects, other arthropods; sugar water from feeders.
HABITAT Old-growth forest clearings, bushy country, and urban gardens while breeding.
LENGTH 3½in (9cm)
WINGSPAN 5in (13cm)

Calliope Hummingbird (S)

Stellula calliope

The Calliope Hummingbird is North America's smallest bird. Despite its diminutive size, it is fierce; the females even attack squirrels trying to rob their nests. Males are distinctive, but the plainer females can be confused with other hummingbird species when their small size is not evident for comparison. The male courtship display includes J-shaped dives, and a buzzing hover display in front of a female.

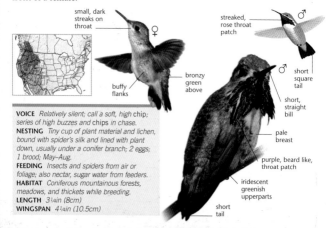

small, dark streaks on throat

♀

bronzy green above

buffy flanks

streaked, rose throat patch

♂

♂ short square tail

short, straight bill

pale breast

purple, beard like, throat patch

iridescent greenish upperparts

short tail

VOICE Relatively silent; call a soft, high chip; series of high buzzes and chips in chase.
NESTING Tiny cup of plant material and lichen, bound with spider's silk and lined with plant down, usually under a conifer branch; 2 eggs; 1 brood; May–Aug.
FEEDING Insects and spiders from air or foliage; also nectar, sugar water from feeders.
HABITAT Coniferous mountainous forests, meadows, and thickets while breeding.
LENGTH 3¼in (8cm)
WINGSPAN 4¼in (10.5cm)

Kingfishers

Kingfishers are primarily a tropical family. There are approximately 90 species of kingfisher in the world, but only one, the Belted Kingfisher, is widespread and found in Canada. Although lacking the array of bright blues, greens, and reds associated with their tropical and European counterparts, Belted Kingfishers are striking birds, distinguished by shaggy head crests, breastbands, and white underparts. The females of the species are more brightly colored than the males, sporting chestnut-colored breastbands. Belted Kingfishers are primarily fish-eaters. They will perch along a waterway for hunting, or hover if no perch is available, and will plunge headfirst into the water to catch their prey. They routinely stun their prey by beating it against a perch before turning the fish around so that it can be eaten head first.

FISH DINNER
A female Belted Kingfisher, showing off its chestnut-colored breastband, uses its large bill to catch and hold slippery prey.

Belted Kingfisher

S

Megaceryle alcyon

Its stocky body, double-pointed crest, large head, and contrasting white collar distinguish the Belted Kingfisher from other species in its range. This kingfisher's loud and far-carrying rattles are heard more often than the bird is seen. Interestingly, it is one of the few birds in North America in which the female is more colorful than the male. The Belted Kingfisher can be found in a large variety of aquatic habitats, both coastal and inland, vigorously defending its territory, all year round.

prominent crest

♀

chestnut band across breast

chestnut flanks

barred tail

large head

bluish gray head with shaggy crest

single blue breastband

♂

long, thick, powerful bill

white collar

double crest

bluish slate upperparts

white collar

white belly

♂

single dark breastband

● ♂

VOICE *Harsh mechanical rattle like an "angry chatter" in flight or from a perch; screams or trill-like warble during breeding.*
NESTING *Unlined chamber in subterranean burrow 3–6ft (1–2m) deep, in earthen bank usually over water; 6–7 eggs; 1 brood; Mar–Jul.*
FEEDING *Plunge-dives from perch to catch fish near the surface, including sticklebacks and trout; also crustaceans, such as crayfish.*
HABITAT *Clear, open waters of streams, rivers, lakes, and estuaries where perches are available.*
LENGTH *11–14in (28–35cm)*
WINGSPAN *19–23in (48–58cm)*

Woodpeckers

Woodpeckers, sapsuckers, and flickers are adapted to living on tree trunks.

The pecking and drumming of woodpeckers, used to construct nest cavities and communicate, is made possible by special skull and beak features, adapted to withstand the shock of continual impact.

Sapsuckers feed on tree sap as a primary source of nourishment. They have tongues tipped with stiff hairs to allow sap to stick to them. The holes the birds create in order to extract the sap also attract insects, which make up the main protein source in the sapsucker diet.

Flickers spend more time feeding on the ground than other woodpeckers, consuming ants and other insects, often foraging in open areas around human habitation. They are notable for their colorful underwing feathers and their distinctive white rump.

COMMON FLICKER
The striking Northern Flicker can be found across the entire North American continent.

Lewis' Woodpecker

Melanerpes lewis

The iridescent dark green back and the salmon-red abdomen of Lewis' Woodpecker distinguish it from any other bird in North America. Juveniles, however, have a brownish head and underparts and lack the gray collar, red face, and pink belly. Lewis' Woodpecker is also notably quieter than other woodpeckers, but it aggressively defends its food sources from other woodpeckers, especially in the winter.

long, broad wings

blackish green rump

very dark upperparts

dull brown head

no gray collar

blackish brown upperparts

blackish green head

black bill

broad silvery gray collar

dark red forehead and cheek

silvery and rosy red underparts

black upperparts with glossy green sheen

blackish green tail

VOICE *Churrs sound; drumming not loud.*
NESTING *Cavity nester, usually in dead tree trunks, with preference for natural cavities and previously used nest holes; 6–7 eggs; 1 brood; May–Aug.*
FEEDING *Flying insects while breeding; acorns, other nuts, and fruits at other times.*
HABITAT *Open Ponderosa Pine forests while breeding, especially old growth modified by burning; riverside woodlands.*
LENGTH 10–11in (25–28cm)
WINGSPAN 19–20in (48–51cm)

Red-headed Woodpecker

Melanerpes erythrocephalus

The Red-headed Woodpecker is easy to identify with its completely red head. Unlike most other woodpecker species, it forages for food—both insects and nuts—and stores it for eating at a later time. It is one of the most skilled flycatchers in the woodpecker family. Its numbers have declined, largely because of the destruction of its habitat.

red head

white rump

bluish gray bill

brownish head

bright red hood

narrow black "necklace"

wing feathers white with black barring

upperparts black with bluish sheen

white wing feathers

VOICE *Extremely harsh, loud churr; no song; active drummer.*
NESTING *Excavates cavity in dead wood; 3–5 eggs; 1–2 broods; May–Aug.*
FEEDING *Forages in flight, on ground, and in trees; insects, spiders, nuts seeds, berries, and fruit; rarely, small mammals such as mice.*
HABITAT *Open deciduous woodlands, including riverine areas, orchards, parks, and forest edges.*
LENGTH 8½–9½in (22–24cm)
WINGSPAN 16–18in (41–46cm)

Red-bellied Woodpecker

Melanerpes carolinus

This woodpecker has expanded its range northward into Canada and westward in the last decade or two. It does not actually possess a red belly; the male is distinguished by its red forehead, crown, and nape, while the female only has a red nape. Males excavate several holes in trees, one of which the female chooses for a nest.

white patches at base of outer wing

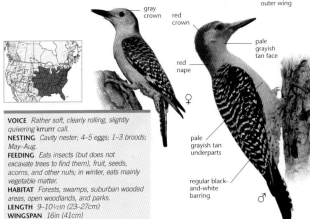

gray crown

red crown

red nape

pale grayish tan face

♀

pale grayish tan underparts

regular black-and-white barring

♂

VOICE *Rather soft, clearly rolling, slightly quivering krrurrr call.*
NESTING *Cavity nester; 4–5 eggs; 1–3 broods; May–Aug.*
FEEDING *Eats insects (but does not excavate trees to find them), fruit, seeds, acorns, and other nuts; in winter, eats mainly vegetable matter.*
HABITAT *Forests, swamps, suburban wooded areas, open woodlands, and parks.*
LENGTH *9–10½in (23–27cm)*
WINGSPAN *16in (41cm)*

Williamson's Sapsucker

Sphyrapicus thyroideus

Unlike other sapsuckers, the male and female plumages of Williamson's Sapsuckers are very dissimilar. The species has very specific habitat needs, partly because of its dependence on the sap and phloem, the innermost bark layer of trees. It can be located in the breeding season by its rather hesitant drumming, which occurs in an uneven series.

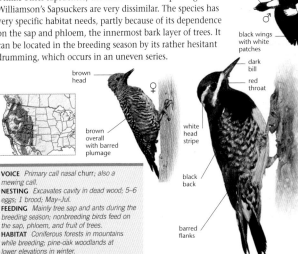

black tail

♂

black wings with white patches

dark bill

red throat

white head stripe

black back

brown head

♀

brown overall with barred plumage

barred flanks

♂

VOICE *Primary call nasal churr; also a mewing call.*
NESTING *Excavates cavity in dead wood; 5–6 eggs; 1 brood; May–Jul.*
FEEDING *Mainly tree sap and ants during the breeding season; nonbreeding birds feed on the sap, phloem, and fruit of trees.*
HABITAT *Coniferous forests in mountains while breeding; pine-oak woodlands at lower elevations in winter.*
LENGTH *9in (23cm)*
WINGSPAN *17in (43cm)*

Yellow-bellied Sapsucker

Ⓢ

Sphyrapicus varius

The Yellow-bellied Sapsucker has distinctive red, black, and white coloring and a soft yellow wash on its underparts. With its relatives, the Red-breasted Sapsucker and the Red-naped Sapsucker, it shares the habit of drilling holes in trees to drink sap. This is the only North American woodpecker that is completely migratory, with females moving farther south than males.

red forehead

red throat

pale yellow on breast and belly

white patch on inner wing

white rump

♂

black-and-white patterned face

dark brown forehead

no red on throat

black-and-white barring on back

white throat

♀

♂

VOICE *Primary call a mewing wheer-wheer-wheer.*
NESTING *Cavities in dead trees; 5–6 eggs; 1 brood; May–Jun.*
FEEDING *Drills holes in trees for sap; ants and other small insects; inner bark of trees, also a variety of fruit; occasionally suet feeders.*
HABITAT *Deciduous forests or mixed deciduous-coniferous forests while breeding; prefers young forests; open wooded areas in winter.*
LENGTH *8–9in (20–23cm)*
WINGSPAN *16–18in (41–46cm)*

Red-naped Sapsucker

Sphyrapicus nuchalis

The Red-naped Sapsucker is closely related to the Red-breasted Sapsucker and the Yellow-bellied Sapsucker. Where these three species overlap geographically, they occasionally interbreed, and birds with intermediate plumage can sometimes be seen. Like the other sapsuckers, this bird drills concentric rings in trees, and extends its specialized tongue to reach the sap.

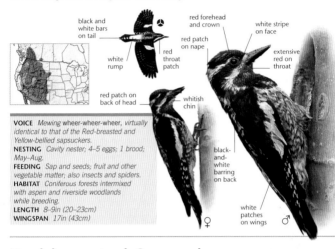

VOICE *Mewing wheer-wheer-wheer, virtually identical to that of the Red-breasted and Yellow-bellied sapsuckers.*
NESTING *Cavity nester; 4–5 eggs; 1 brood; May–Aug.*
FEEDING *Sap and seeds; fruit and other vegetable matter; also insects and spiders.*
HABITAT *Coniferous forests intermixed with aspen and riverside woodlands while breeding.*
LENGTH *8–9in (20–23cm)*
WINGSPAN *17in (43cm)*

Red-breasted Sapsucker

Sphyrapicus ruber

Apart from its distinctive red head and breast, the Red-breasted Sapsucker resembles other sapsuckers. Like its relatives, it drills holes in tree trunks to extract sap. Other birds and mammals, such as squirrels and bats, obtain food from these holes. The northern form, *S. r. ruber*, occurs from Alaska to Oregon, and has a back lightly marked with gold spots and a brightly colored head.

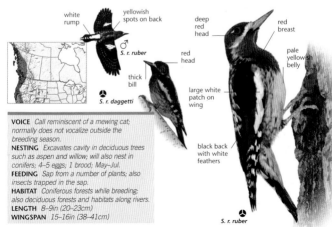

VOICE *Call reminiscent of a mewing cat; normally does not vocalize outside the breeding season.*
NESTING *Excavates cavity in deciduous trees such as aspen and willow; will also nest in conifers; 4–5 eggs; 1 brood; May–Jul.*
FEEDING *Sap from a number of plants; also insects trapped in the sap.*
HABITAT *Coniferous forests while breeding; also deciduous forests and habitats along rivers.*
LENGTH *8–9in (20–23cm)*
WINGSPAN *15–16in (38–41cm)*

Downy Woodpecker ⓢ

Picoides pubescens

The smallest North American woodpecker, the Downy Woodpecker is seen all year round from coast to coast in Canada. This woodpecker is distinguished from the similar Hairy Woodpecker by its shorter bill and much smaller size. After breeding, Downy Woodpeckers remain in the same area, but wander around in search of food in a variety of habitats, including suburbs and gardens.

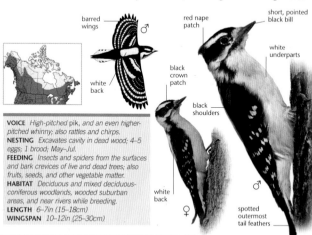

barred wings ♂

white back

red nape patch

short, pointed black bill

white underparts

black crown patch

black shoulders

white back ♀

black ♂

spotted outermost tail feathers

VOICE High-pitched pik, and an even higher-pitched whinny; also rattles and chirps.
NESTING Excavates cavity in dead wood; 4–5 eggs; 1 brood; May–Jul.
FEEDING Insects and spiders from the surfaces and bark crevices of live and dead trees; also fruits, seeds, and other vegetable matter.
HABITAT Deciduous and mixed deciduous-coniferous woodlands, wooded suburban areas, and near rivers while breeding.
LENGTH 6–7in (15–18cm)
WINGSPAN 10–12in (25–30cm)

Hairy Woodpecker ⓢ

Picoides villosus

Like the Downy Woodpecker, the Hairy Woodpecker is widespread in North America, breeding and wintering from coast to coast in the US and Canada. The two species look quite similar, but the Hairy Woodpecker has a larger, thicker bill and is almost twice as long in size. It prefers forests, where it uses live tree trunks both as nesting sites and as places to forage.

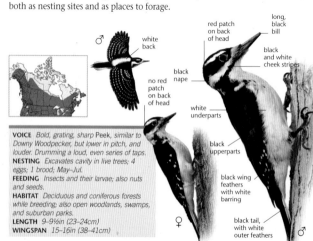

♂

white back

red patch on back of head

long, black bill

black and white cheek stripes

black nape

no red patch on back of head

white underparts

black upperparts

black wing feathers with white barring

♀

black tail, with white outer feathers ♂

VOICE Bold, grating, sharp Peek, similar to Downy Woodpecker, but lower in pitch, and louder. Drumming a loud, even series of taps.
NESTING Excavates cavity in live trees; 4 eggs; 1 brood; May–Jul.
FEEDING Insects and their larvae; also nuts and seeds.
HABITAT Deciduous and coniferous forests while breeding; also open woodlands, swamps, and suburban parks.
LENGTH 9–9½in (23–24cm)
WINGSPAN 15–16in (38–41cm)

White-headed Woodpecker

Picoides albolarvatus

The White-headed Woodpecker's plumage pattern is unique among North American woodpeckers. While it is common in some areas of its geographically restricted range, its population is vulnerable because of forest fragmentation. Ponderosa Pine seeds are basic to its diet, and poor pine crops may result in low breeding success. In winter, males and females tend to forage separately, and females feed lower in trees.

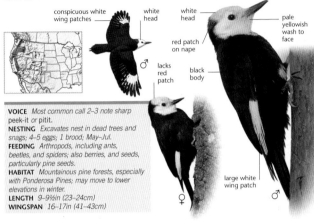

conspicuous white wing patches

white head

white head

red patch on nape

pale yellowish wash to face

♂

lacks red patch

black body

VOICE *Most common call 2–3 note sharp peek-it or pitit.*
NESTING *Excavates nest in dead trees and snags; 4–5 eggs; 1 brood; May–Jul.*
FEEDING *Arthropods, including ants, beetles, and spiders; also berries, and seeds, particularly pine seeds.*
HABITAT *Mountainous pine forests, especially with Ponderosa Pines; may move to lower elevations in winter.*
LENGTH *9–9½in (23–24cm)*
WINGSPAN *16–17in (41–43cm)*

large white wing patch

♂

♀

American Three-toed Woodpecker

Picoides dorsalis

This species breeds farther north than any other North American woodpecker, including the Black-backed Woodpecker. It resembles the Black-backed Woodpecker in terms of size, head markings, and because they are the only two North American woodpeckers with three toes on each foot. These two species require matures forests with old or dead trees.

black-and-white barred back

♂

large yellow patch

black head

long, straight bill

black-and-white streaked crown

white breast

VOICE *Queep, quip, or pik; generally quiet, likened to the Yellow-bellied Sapsucker.*
NESTING *Excavates cavity mainly in dead or dying wood, sometimes in live wood; 4 eggs; 1 brood; May–Jul.*
FEEDING *Flakes off bark and eats insects underneath, mainly larvae of Bark Beetles.*
HABITAT *Mature, northerly coniferous forests; largely nonmigratory; more open areas in winter.*
LENGTH *8–9in (20–23cm)*
WINGSPAN *15in (38cm)*

black bars on flanks

black, slightly forked tail, with white outer tail feathers

♀

♂

Black-backed Woodpecker

S

Picoides arcticus

Despite being widespread, this bird is difficult to find. It often occurs in areas of burned forest, eating wood-boring beetles that occur after outbreaks of fire. This diet is very restrictive, and the species is greatly affected by the prevention of forest fires. Although its range overlaps with the American Three-toed Woodpecker, the two are rarely found together in the same locality.

black back

♀

white spots on outer wings

black cap

♂

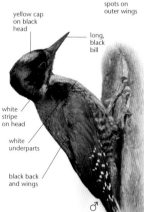

yellow cap on black head

long, black bill

white stripe on head

white underparts

black back and wings

♂

VOICE *Single pik.*
NESTING *Cavity excavated in tree; 3–4 eggs; 1 brood; May–Jul.*
FEEDING *Flakes off bark and eats beetles underneath, especially larvae of wood-boring beetles.*
HABITAT *Northerly, mountainous coniferous forests that require fire for renewal; moves from place to place following outbreaks of wood-boring beetles.*
LENGTH 9–9½in (23–24cm)
WINGSPAN 15–16in (38–41cm)

Northern Flicker

S

Colaptes auratus

The Northern Flicker is a ground forager. The two subspecies, the Yellow-shafted Flicker in the East, and Red-shafted Flicker in the West, interbreed in a wide area in the Great Plains. They can be distinguished when in flight, as the underwing feathers will either be a vivid yellow or a striking red, as their names indicate.

black crescent

♂
RED-SHAFTED FORM

orangish red underwings

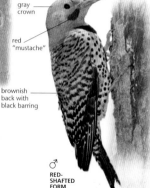

no "mustache"

♀
YELLOW-SHAFTED FORM

buffy forehead

gray crown

red "mustache"

brownish back with black barring

♂
RED-SHAFTED FORM

VOICE *Loud kew-kew-kew, each note ascending at the end; softer wicka-wicka-wicka.*
NESTING *Cavity usually in dead wood, but sometimes in live wood; 6–8 eggs; 1 brood; May–Jun.*
FEEDING *Mainly ants in breeding season; also fruits in winter.*
HABITAT *Open woodlands and forest edges while breeding; also suburbs.*
LENGTH 12–13in (31–33cm)
WINGSPAN 19–21in (48–53cm)

Pileated Woodpecker

(S)

Dryocopus pileatus

The largest woodpecker in North America, the Pileated Woodpecker is instantly recognizable by its spectacular large, tapering, bright-red crest. A mated pair of Pileated Woodpeckers defends their breeding territory all year. Indeed, a pair may live in the same old, dead tree every year, but will hammer out a new nest cavity with their powerful bills each season. Their rectangular-shaped, abandoned nest cavities are sometimes reused by other birds, and occasionally inhabited by mammals.

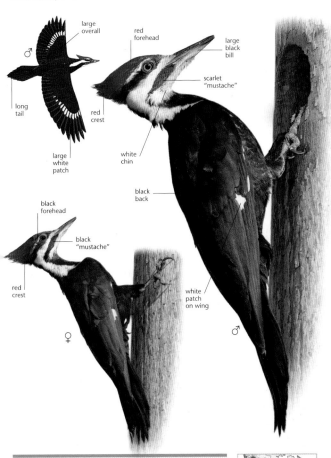

large overall

red forehead

large black bill

scarlet "mustache"

♂

long tail

red crest

large white patch

white chin

black back

black forehead

black "mustache"

red crest

white patch on wing

♀

♂

VOICE *High-pitched, quite loud yuck-yuck-yuck and yuka-yuka-yuka.*
NESTING *Excavates rectangular-shaped cavity, usually in dead tree; 3–5 eggs; 1 brood; May–July.*
FEEDING *Bores deep into trees and peels off large strips of bark for carpenter ants and beetle larvae; also digs on ground and on fallen logs, and opportunistically eats fruit and nuts.*
HABITAT *Deciduous and coniferous forests and woodlands while breeding; also swampy areas*
LENGTH *16–18in (41–46cm)*
WINGSPAN *26–30in (66–76cm)*

Flycatchers

The North American species of flycatchers are all members of a single family—the Tyrant Flycatchers. They are uniform in appearance; most are drab colored, olive-green or gray birds, sometimes with yellow on the underparts. The members of the genus *Empidonax* include some of the most difficult birds to identify in North America; they are best distinguished by their songs. Typical flycatcher feeding behavior is to sit on a branch or exposed perch sallying forth to catch flying insects. Many live in wooded habitats, though the kingbirds prefer woodland edges and deserts. Nearly all flycatchers are long-distance migrants and spend the winter in Central and South America.

BIG MOUTHS
Young Dusky Flycatchers display the wide bills that help them to catch flying insects as adults.

Eastern Phoebe

S

Sayornis phoebe

The Eastern Phoebe is an early spring migrant that tends to nest under bridges, culverts, and on buildings, in addition to rocky outcroppings. Not shy, it is also familiar because of its *fee-bee* vocalization and constant tail wagging. By tying a thread on the leg of several Eastern Phoebes, ornithologist John James Audubon established that individuals return from the south to a previously used nest site. Although difficult to tell apart, males tend to be slightly larger and darker than females.

rounded wings with two faint wing bars

brownish gray upperparts

white throat

pale edges to wing feathers

white throat

♂

round, dark-capped head

yellowish tint on lower belly

dark eye

long, dark tail

olive tint to sides and breast

🔵 BREEDING

VOICE *Common call a clear, weak chip; song an emphatic* fee-bee *or* fee-b-be-bee.
NESTING *Open cup of mud, moss, and leaves, almost exclusively on manmade structures; 3–5 eggs; 2 broods; Apr–Jul.*
FEEDING *Flying insects such as wasps, flies, moths, and dragonflies; also small fruits from fall through winter.*
HABITAT *Open woodland and along deciduous or mixed forest edges, gardens and parks near water.*
LENGTH *5½–7in (14–17cm)*
WINGSPAN *10½in (27cm)*

Say's Phoebe

S

Sayornis saya

Say's Phoebe breeds farther north than any other flycatcher in its family. It is not particularly shy around people, and from early spring to late fall it is a common sight on ranches and farms. Its contrasting dark cap is conspicuous as it perches on bushes or power lines, often wagging its tail. Shortly after a pair is formed, the male will hover in front of potential nest sites.

dark eye

sooty gray cap

faint wing bars

small, black bill

grayish brown back

black tail

pale sooty gray neck and breast

black feet and legs

rufous undertail and lower belly

VOICE Pee-ee *or* pee-ur; *also whistled chur-eep, which may be integrated with a chatter; primary song a pit-see-eur and* pit-eet.
NESTING *Shallow cup of twigs, moss, or stems on ledge or in rocky crevice; 3–7 eggs; 1–2 broods; Apr–Jul.*
FEEDING *Insects in flight, such as beetles, wasps, grasshoppers, and crickets; berries.*
HABITAT *Dry, open country (agricultural areas); generally avoids watercourses.*
LENGTH *7in (17.5cm)*
WINGSPAN *13in (33cm)*

Olive-sided Flycatcher

T

Contopus cooperi

The Olive-sided Flycatcher is identified by its distinctive song, large size, and contrasting belly and flank colors, which make its underside appear like a vest with the buttons undone. Breeding pairs are known to aggressively defend their territory. This flycatcher undertakes a long journey from northern parts of North America to winter in Panama and the Andes.

short tail

large, dark head

lower base of bill often dull orange

dull white throat

brownish gray back

pointed wings

white belly

brownish olive flanks

VOICE *Even pip-pip-pip; loud 3-note whistle:* quick-THREE-BEERS *or* whip-WEE-DEER.
NESTING *Open cup of twigs, rootlets, lichens; 2–5 eggs; 1 brood; May–Aug.*
FEEDING *Waits on perch before swooping after prey; flying insects, such as bees, wasps, and flying ants.*
HABITAT *Mountainous coniferous forests at edges or near ponds, bogs, meadows with standing dead trees; suburban areas.*
LENGTH *7–8in (18–20cm)*
WINGSPAN *13in (33cm)*

Western Wood-pewee

Contopus sordidulus

Found in many forested habitats of western North America, the Western Wood-pewee vocalizes from high perches, principally during the breeding season, but also during winter and migration. It forages aerially on insects in a similar fashion to swallows. Adults are aggressive toward laying parasitic intruders; however, they accept Brown-headed Cowbird eggs, though few fledge successfully.

pale throat

pointed wings

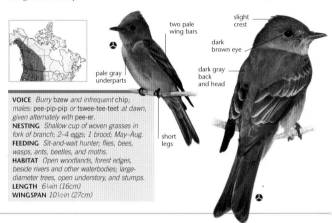

two pale wing bars

slight crest

dark brown eye

dark gray back and head

pale gray underparts

VOICE *Burry* bzew *and infrequent* chip; *males:* pee-pip-pip *or* tswee-tee-teet *at dawn, given alternately with* pee-er.
NESTING *Shallow cup of woven grasses in fork of branch; 2–4 eggs; 1 brood; May–Aug.*
FEEDING *Sit-and-wait hunter; flies, bees, wasps, ants, beetles, and moths.*
HABITAT *Open woodlands, forest edges, beside rivers and other waterbodies; large-diameter trees, open understory, and stumps.*
LENGTH *6¼in (16cm)*
WINGSPAN *10½in (27cm)*

short legs

Eastern Wood-pewee

Contopus virens

The Eastern Wood-pewee is found in many types of woodland in southern and eastern Canada. The male is slightly larger than the female, but their plumage is practically identical. Recent population declines have been attributed to loss of breeding habitat due to heavy browsing by White-tailed Deer. This has been compounded by the Eastern Wood-pewee's susceptibility to brood parasitism by Brown-headed Cowbirds.

yellow lower mandible

slightly ragged crest

partial eye-ring

pale throat

pale gray

thin, white wing bars

pointed wings

yellowish wash on underparts

thin, white edges to wing feathers

VOICE *Terse* chip; *slurred* pee-ah-wee *or* ah di dee, *plaintive* wee-ooo, *or* wee-ur.
NESTING *Shallow cup of grass, lichens on horizontal limb; 2–4 eggs; 1 brood; May–Sep.*
FEEDING *Mainly flying insects, such as flies, beetles, and bees; occasionally forages for insects on foliage on the ground.*
HABITAT *Deciduous and coniferous forests near clearing or edges while breeding; near watersides in Midwest.*
LENGTH *6in (15cm)*
WINGSPAN *9–10in (23–26cm)*

Yellow-bellied Flycatcher

S

Empidonax flaviventris

The Yellow-bellied Flycatcher is characteristic of northern coniferous forests and Sphagnum-moss peatlands. It is not well known, because of its remote habitats and its secretive habits; it is much more often heard than seen. It remains on its breeding grounds for only about two months, then migrates to winter quarters in southern Mexico and Central America to Panama.

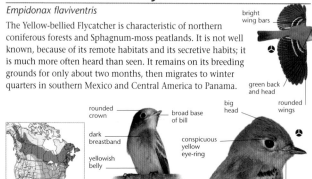

bright wing bars

green back and head

rounded wings

big head

rounded crown

broad base of bill

dark breastband

yellowish belly

conspicuous yellow eye-ring

yellow-olive throat

white wing bars

short, narrow, square tail

VOICE *Chu-wee and abrupt brrrrt; abrupt killink, che-lek, or che-bunk, with variations.*
NESTING *Cup of moss, twigs, and needles on or near ground, often in a bog; 3–5 eggs; 1 brood; Jun–Jul.*
FEEDING *Insects in the air or mosquitoes, midges, and flies from foliage; sometimes berries and seeds.*
HABITAT *Boreal forests and bogs with spruce trees while breeding.*
LENGTH *5½in (14cm)*
WINGSPAN *8in (20cm)*

Acadian Flycatcher

Empidonax virescens

Its often-drooped wings and minimal wing and tail flicking give the Acadian Flycatcher an outwardly calm appearance compared to other flycatchers. It bathes by diving into water, then preens on a perch. It suffers more parasitism from Brown-headed Cowbirds in small woodlots than in large forests. Where Cowbirds lay their eggs in the flycatcher's nest, they displace the flycatcher's young.

prominent wing bars

slight crest

narrow, eye-ring

greenish nape and back

two wing bars

broad bill with yellowish lower mandible

pale belly

broad tail

yellowish wash on lower belly

white-edged flight feathers

VOICE *Soft peet, one of many calls; territorial song tee-chup, peet-sah, or flee-sick, loud and "explosive" sounding.*
NESTING *Shallow, open cup in tree fork or shrub near water; 3 eggs; 2 broods; May–Aug.*
FEEDING *Insects from undersides of leaves, in the air, and occasionally on the ground; also berries.*
HABITAT *Mature deciduous forests near water, preferably undisturbed, while breeding.*
LENGTH *6in (15cm)*
WINGSPAN *9in (23cm)*

Willow Flycatcher (S)

Empidonax traillii

The Willow Flycatcher is only distinguished from the nearly identical Alder Flycatcher by its song. A territorial bird, it spreads its tail and flicks it upward during aggressive encounters. It is, however, frequently parasitized by the Brown-headed Cowbird, which lays its eggs in the flycatcher's nest and removes its host's eggs.

two buff to yellow wing bars

square tail

grayish green upperparts

brown eye

thin eye-ring

dark upper mandible

paler lower mandible

yellow-tinged flanks

whitish belly

dark legs and toes

dark tail

VOICE Soft, dry whit and several buzzy notes; sharp fitz-bew with accent on the first syllable; also creet.
NESTING Rather loose, untidy cup in base of shrub near water; 3–4 eggs; 1 brood; May–Aug.
FEEDING Insects, mostly caught in flight; fruit in winter.
HABITAT Willow thickets and other moist shrubby areas along watercourses while breeding.
LENGTH 5–6¾in (13–17cm)
WINGSPAN 7½–9½in (19–24cm)

Alder Flycatcher (S)

Empidonax alnorum

Until 1973 the Alder Flycatcher and the Willow Flycatcher were considered to be one species called Traill's Flycatcher. The two species cannot be reliably identified by sight, but they do have distinctive songs. The Alder Flycatcher also breeds farther north than the Willow Flycatcher, arriving late in spring and leaving early in fall. Its nests are extremely hard to locate.

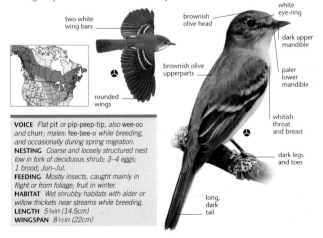

two white wing bars

brownish olive head

white eye-ring

dark upper mandible

paler lower mandible

brownish olive upperparts

rounded wings

whitish throat and breast

dark legs and toes

long, dark tail

VOICE Flat pit or pip-peep-tip, also wee-oo and churr; males: fee-bee-o while breeding, and occasionally during spring migration.
NESTING Coarse and loosely structured nest low in fork of deciduous shrub; 3–4 eggs; 1 brood; Jun–Jul.
FEEDING Mostly insects, caught mainly in flight or from foliage; fruit in winter.
HABITAT Wet shrubby habitats with alder or willow thickets near streams while breeding.
LENGTH 5¾in (14.5cm)
WINGSPAN 8½in (22cm)

Least Flycatcher

(S)

Empidonax minimus

The Least Flycatcher is a solitary bird and is very aggressive towards intruders encroaching upon its breeding territory. This combative behavior reduces the likelihood of brood parasitism by the Brown-headed Cowbird. The Least Flycatcher is very active, and frequently flicks its wings and tail upward. Common across Canada in mixed and deciduous woodlands, especially at the edges, it spends a short time—up to only two months—on its northern breeding grounds before migrating south. Adults molt in winter, while young molt before and during fall migration.

short, broad-based bill

large head

short, narrow tail

two wing bars

pale throat

buffy wing bars

marked, white eye-ring

greenish brown back

short wings

pale yellow belly

VOICE *Soft, short whit; frequent, persistent, characteristic tchebeck, sings during spring migration and breeding season.*
NESTING *Compact cup of tightly woven bark strips and plant fibers in fork of deciduous tree; 3–5 eggs; 1 brood; May–Jul.*
FEEDING *Insects, such as flies, midges, beetles, ants, butterflies, and larvae; occasionally berries and seeds.*
HABITAT *Coniferous and mixed deciduous forests while breeding; conifer groves or wooded wetlands.*
LENGTH *5¼in (13.5cm)*
WINGSPAN *7¾in (19.5cm)*

Hammond's Flycatcher

⒮

Empidonax hammondii

Hammond's Flycatcher is a migrant from Central and South America. Its distinctive song is mostly heard on its breeding grounds. In the breeding season, males are competitive and aggressive, and are known to lock together in mid-air to resolve their territorial squabbles. Since this species is dependent on mature old-growth forest, logging is thought to be adversely affecting its numbers.

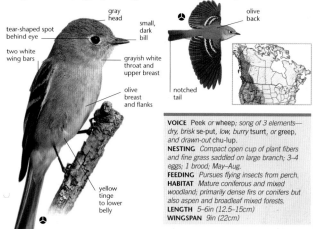

gray head

small, dark bill

tear-shaped spot behind eye

two white wing bars

grayish white throat and upper breast

olive breast and flanks

yellow tinge to lower belly

olive back

notched tail

VOICE *Peek* or *wheep*; song of 3 elements— dry, brisk *se-put*, low, burry *tsurrt*, or *greep*, and drawn-out *chu-lup*.
NESTING *Compact open cup of plant fibers and fine grass saddled on large branch; 3–4 eggs; 1 brood; May–Aug.*
FEEDING *Pursues flying insects from perch.*
HABITAT *Mature coniferous and mixed woodland; primarily dense firs or conifers but also aspen and broadleaf mixed forests.*
LENGTH *5–6in (12.5–15cm)*
WINGSPAN *9in (22cm)*

Dusky Flycatcher

⒮

Empidonax oberholseri

The Dusky Flycatcher waits on a perch for a flying insect, flies out to catch it, and then returns to its position to consume it, often wiping its bill on the perch after completing its meal. It lives in mountainous areas of western Canada and the US, where it is vulnerable to storms that can severely impact a local breeding population by flattening the trees.

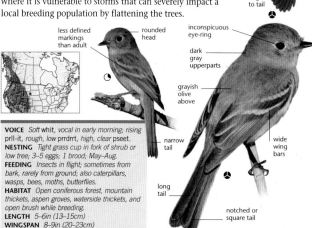

wing bars

faint, white edge to tail

less defined markings than adult

rounded head

inconspicuous eye-ring

dark gray upperparts

grayish olive above

narrow tail

wide wing bars

long tail

notched or square tail

VOICE *Soft* whit, vocal in early morning; rising *prll-it*, rough, low *prrdrrt*, high, clear *pseet*.
NESTING *Tight grass cup in fork of shrub or low tree; 3–5 eggs; 1 brood; May–Aug.*
FEEDING *Insects in flight; sometimes from bark, rarely from ground; also caterpillars, wasps, bees, moths, butterflies.*
HABITAT *Open coniferous forest, mountain thickets, aspen groves, waterside thickets, and open brush while breeding.*
LENGTH *5–6in (13–15cm)*
WINGSPAN *8–9in (20–23cm)*

Pacific-slope Flycatcher ⓢ

Empidonax difficilis

The Pacific-slope Flycatcher is virtually identical to the
Cordilleran Flycatcher—both were formerly considered to
be one species called the Western Flycatcher. The Pacific-
slope Flycatcher is a short-distance migrant that winters in
Mexico. The female is active during nest-building and
incubation, but the male provides food for nestlings.

yellow-washed throat

rounded wings

yellow-orange lower mandible

slight crest

tear-shaped eye-ring extending behind eye

brown-washed breast

olive back and head

VOICE Chrrip, seet, zeet; *squeaky, repeated ps-*
SEET, ptsick, seet, or TSEE-wee, pttuck, tseep.
NESTING *Open cup in fork of tree, shelf on*
bank, or bridge; 2–4 eggs; 2 broods; Apr–Jul.
FEEDING *Insects caught in air or gleaned from*
foliage: beetles, wasps, bees, flies, moths,
caterpillars, spiders; rarely berries.
HABITAT *Humid coastal coniferous forest,*
Pine-Oak forest, and dense second-growth
forest while breeding.
LENGTH *6–7in (15–17.5cm)*
WINGSPAN *8–9in (20–23cm)*

Cordilleran Flycatcher ⓢ

Empidonax occidentalis

The Cordilleran Flycatcher differs slighty from the Pacific-slope Flycatcher with its
darker upperparts and more olive and yellow underparts. This bird is found east of
the Rocky Mountains, just reaching southeastern British Columbia and
southwestern Alberta. The sexes look alike and are monogamous, behaving
territorially while breeding. Most molting occurs when the birds are wintering.

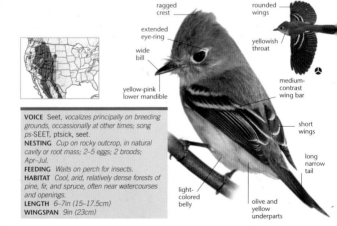

ragged crest

rounded wings

extended eye-ring

yellowish throat

wide bill

yellow-pink lower mandible

medium-contrast wing bar

short wings

long narrow tail

light-colored belly

olive and yellow underparts

VOICE Seet, *vocalizes principally on breeding*
grounds, occasionally at other times; song
ps-SEET, ptsick, seet.
NESTING *Cup on rocky outcrop, in natural*
cavity or root mass; 2–5 eggs; 2 broods;
Apr–Jul.
FEEDING *Waits on perch for insects.*
HABITAT *Cool, arid, relatively dense forests of*
pine, fir, and spruce, often near watercourses
and openings.
LENGTH *6–7in (15–17.5cm)*
WINGSPAN *9in (23cm)*

Western Kingbird

S

Tyrannus verticalis

A conspicuous summer breeder in the lower parts of the western provinces and the western US, the Western Kingbird occurs in open habitats. The white outer edges on its outer tail feathers distinguish it from other kingbirds. A large, loosely defined territory is defended against other kingbirds when breeding begins in spring; a smaller core area is defended as the season progresses.

white-edged tail

olive-gray back

strong, dark eye-line

small bill

white chin

dark wing with no wing bars

gray chest

yellow belly

gray head

gray back

white edge to outer tail feathers

notched tail

VOICE *Whit, pwee-t, and chatter; regularly repeated sharp kip notes and high-pitched notes.*
NESTING *Open, bulky cup of grass, rootlets, and twigs in tree, shrub, utility pole; 2–7 eggs; 1 brood; Apr–Jul.*
FEEDING *Insects and fruit.*
HABITAT *Grasslands, savannah, desert shrub, pastures, and cropland near elevated perches and water.*
LENGTH *8–9in (20–23cm)*
WINGSPAN *15–16in (38–41cm)*

Eastern Kingbird

S

Tyrannus tyrannus

The Eastern Kingbird is a tame yet highly territorial species and is known for its aggressive behavior toward potential predators, particularly crows and hawks. It is able to identify and remove the eggs of the Brown-headed Cowbird when they are laid in its nest. The Eastern Kingbird is generally monogamous and pairs will return to the same territory in subsequent years.

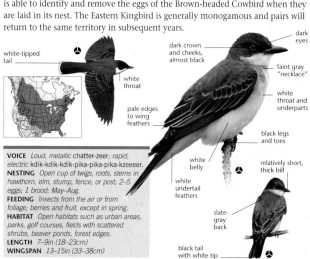

white-tipped tail

white throat

pale edges to wing feathers

dark eyes

dark crown and cheeks, almost black

faint gray "necklace"

white throat and underparts

black legs and toes

relatively short, thick bill

white belly

white undertail feathers

slate-gray back

black tail with white tip

VOICE *Loud, metallic chatter-zeer; rapid, electric kdik-kdik-kdik-pika-pika-pika-kzeeeer.*
NESTING *Open cup of twigs, roots, stems in hawthorn, elm, stump, fence, or post; 2–5 eggs; 1 brood; May–Aug.*
FEEDING *Insects from the air or from foliage; berries and fruit, except in spring.*
HABITAT *Open habitats such as urban areas, parks, golf courses, fields with scattered shrubs, beaver ponds, forest edges.*
LENGTH *7–9in (18–23cm)*
WINGSPAN *13–15in (33–38cm)*

Great Crested Flycatcher

Myiarchus crinitus

The Great Crested Flycatcher is locally common and geographically quite widespread from Alberta and the Maritimes to Florida and Texas, but is often overlooked because it remains in the forest canopy, though it visits the ground for food and nest material. Its presence is usually given away by its loud, sharp, double-syllabled notes. It lines its nest with shed snakeskins (or similar materials like cellophane) like other *Myiarchus* flycatchers.

rusty edges to outer wing feathers

whitish wing bars

gray breast

yellow belly

brown crest

olive-brown back

long, thin bill

gray breast and face

rufous-orange tail

yellow belly

brownish legs and feet

long tail

VOICE *Loud, abrupt* purr-it *given by both sexes; males: repeated* whee-eep, *occasionally* wheeyer.
NESTING *In deep cavity, usually woodpecker hole, lined with leaves, bark, trash, and snakeskins; 4–6 eggs; 1 brood; May–Jul.*
FEEDING *Flying insects, moths, and caterpillars, mainly from leaves and branches in the canopy; also small berries and fruits.*
HABITAT *Deciduous and mixed woodlands with clearings while breeding.*
LENGTH *7–8in (18–20cm)*
WINGSPAN *13in (33cm)*

Vireos

Originally, vireos were associated with warblers, but recent molecular studies suggest that they are actually related to crow-like birds. Vireo plumage is drab, often predominantly greenish or grayish above and whitish below, augmented by eye-rings, "spectacles," eyestripes, and wing bars. They prefer broadleaved habitats, where they hop and climb as they slowly forage for insects and other food.

SOUTHERN BOUND
Like most Vireos, the Blue-headed Vireo migrates south to warmer climates in winters.

Jays, Crows,& Ravens

Collectively known as corvids, jays, crows, and ravens are remarkably social. Always the opportunists, these birds use strong bills and toes to obtain a varied, omnivorous diet. They are among the most intelligent birds, exhibiting self-awareness when looking into mirrors, making tools, and successfully problem-solving.

BLACK AND BLUES
A Steller's Jay shows off its beautiful black, blue, and gray plumage.

Loggerhead Shrike

white flash
in wings

white
edges
to tail

Lanius ludovicianus

Although a songbird, the Loggerhead Shrike is superficially falcon-like in several ways, particularly its prominent black face mask and powerful, hooked bill. It sits atop posts or tall trees, swooping down to catch prey on the ground. It has the unusual habit of impaling its prey on thorns, barbed wire, or sharp twigs, which is thought to be the reason for the nickname "butcher bird."

black
wings

gray
crown

hooked
bill

black
"mask"

pale
undertail
feathers

unstreaked,
gray
underparts

rounded
tail

VOICE *Quiet warbles, trills, and harsh notes singly or in series: chaa chaa chaa.*
NESTING *Open cup of vegetation, placed in thorny tree; 5 eggs; 1 brood; Mar–Jun.*
FEEDING *Kills large insects and small vertebrates—rodents, birds, reptiles—with powerful bill.*
HABITAT *Semi-open country with grazed short grass and scattered perches in generally remote regions; distribution is erratic.*
LENGTH *9in (23cm)*
WINGSPAN *12in (31cm)*

Northern Shrike

Lanius excubitor

conspicuous
white wing bar

pale gray
upperparts

This northern version of the familiar Loggerhead Shrike is an uncommon winter visitor to the northern US and southern Canada. The Northern Shrike is paler, larger bodied, and larger billed than the Loggerhead Shrike, which enables it to attack and subdue larger prey than the latter. Many Northern Shrike populations worldwide are in decline, but to date there is no sign of a similar decline in North America.

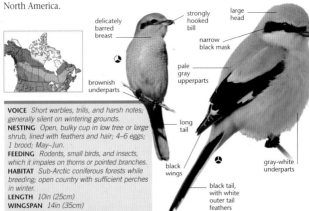

delicately
barred
breast

strongly
hooked
bill

large
head

narrow
black mask

pale
gray
upperparts

brownish
underparts

long
tail

black
wings

black tail,
with white
outer tail
feathers

gray-white
underparts

VOICE *Short warbles, trills, and harsh notes; generally silent on wintering grounds.*
NESTING *Open, bulky cup in low tree or large shrub, lined with feathers and hair; 4–6 eggs; 1 brood; May–Jun.*
FEEDING *Rodents, small birds, and insects, which it impales on thorns or pointed branches.*
HABITAT *Sub-Arctic coniferous forests while breeding; open country with sufficient perches in winter.*
LENGTH *10in (25cm)*
WINGSPAN *14in (35cm)*

White-eyed Vireo

S

Vireo griseus

The White-eyed Vireo is a vocal inhabitant of dense thickets, where it forages actively. It is heard more often than it is seen, singing persistently into the heat of the day and late into the year, long after most birds have become silent. It is parasitized by the Brown-headed Cowbird, and as many as half of the White-eyed Vireo's nestlings do not survive.

short tail

two wing bars

bright yellow "spectacles"

white eye

gray nape

whitish throat

two prominent wing bars

yellow flanks

yellow-and-black wing markings

VOICE *Raspy, angry scold; males: highly variable repertoire of over a dozen distinct songs.*
NESTING *Deep cup in dense vegetation, lined with fine fibers, often near water, suspended from twigs by the rim; 3–5 eggs; 2 broods; Mar–Jul.*
FEEDING *Bees, flies, beetles, and bugs, plucked from leaves or from the air; fruit in winter.*
HABITAT *Dense brush, scrub while breeding.*
LENGTH *5in (13cm)*
WINGSPAN *7½in (19cm)*

Yellow-throated Vireo

S

Vireo flavifrons

This large vireo of southeastern Canada from Manitoba to Quebec and eastern US woodlands is usually found foraging and singing high in the canopy. It has a bright yellow throat, breast, and "spectacles," and a white belly and flanks. The fragmentation of forests, spraying of insecticides, and cowbird parasitism have led to regional declines in Yellow-throated Vireo populations, but the bird's range, as a whole, has expanded.

conspicuous white wing bars

bright, yellow "spectacles" and patch between eye and bill

olive back

gray rump

yellow throat and breast

white belly

fairly short tail

white undertail feathers

VOICE *Scolding, hoarse, rapid calls; males: slow, repetitive, two- or three-note phrase, separated by long pauses.*
NESTING *Rounded cup of fibers bound with spider webs, near the top of a large tree and hung by the rim; 3–5 eggs; 1 brood; Apr–Jul.*
FEEDING *Forages insects from branches high in trees; also fruit when available.*
HABITAT *Extensive, mature, deciduous and mixed woodlands while breeding.*
LENGTH *5½in (14cm)*
WINGSPAN *9½in (24cm)*

Cassin's Vireo ⓢ

Vireo cassinii

Cassin's Vireo is similar to the closely related Blue-headed Vireo in appearance and song. It is conspicuous and vocal throughout its breeding grounds in the far west of the US and north into southwest Canada. In winter, virtually the entire population migrates to Mexico. It was named in honor of John Cassin, who published the first comprehensive study of North American birds in 1865.

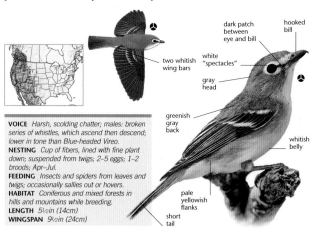

dark patch between eye and bill

hooked bill

two whitish wing bars

white "spectacles"

gray head

greenish gray back

whitish belly

pale yellowish flanks

short tail

VOICE *Harsh, scolding chatter; males: broken series of whistles, which ascend then descend; lower in tone than Blue-headed Vireo.*
NESTING *Cup of fibers, lined with fine plant down; suspended from twigs; 2–5 eggs; 1–2 broods; Apr–Jul.*
FEEDING *Insects and spiders from leaves and twigs; occasionally sallies out or hovers.*
HABITAT *Coniferous and mixed forests in hills and mountains while breeding.*
LENGTH *5½in (14cm)*
WINGSPAN *9½in (24cm)*

Blue-headed Vireo ⓢ

Vireo solitarius

The Blue-headed Vireo has a blue-gray, helmeted head, adorned with striking white "spectacles" around its dark eyes, which helps to distinguish it from other vireos in its range. This stocky and slow moving bird is heard more often than it is seen in its forest breeding habitat. During migration it can be more conspicuous and is the first vireo to return in spring.

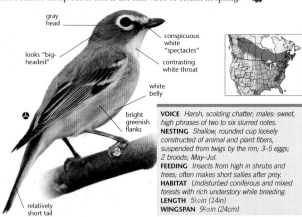

two wing bars

greenish back

gray head

looks "big-headed"

conspicuous white "spectacles"

contrasting white throat

white belly

bright greenish flanks

relatively short tail

VOICE *Harsh, scolding chatter; males: sweet, high phrases of two to six slurred notes.*
NESTING *Shallow, rounded cup loosely constructed of animal and plant fibers, suspended from twigs by the rim; 3–5 eggs; 2 broods; May–Jul.*
FEEDING *Insects from high in shrubs and trees; often makes short sallies after prey.*
HABITAT *Undisturbed coniferous and mixed forests with rich understory while breeding.*
LENGTH *5½in (14in)*
WINGSPAN *9½in (24cm)*

Hutton's Vireo

S

Vireo huttoni

This bird is part of a geographically variable species with about a dozen subspecies, grouped into two populations. The coastal population occurs from British Columbia to Baja California; the second population is found in the Southwest down to Central America, and may actually be a different species. Hutton's Vireo is distinguishable from the Ruby-crowned Kinglet by its larger size and thicker bill. It is largely non-migratory.

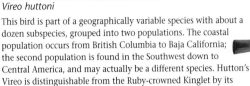

short, rounded wings

pale patch

large, rounded head

broken eye-ring

thick, hooked bill

white wing bars

white wing bars

blue-gray legs

VOICE *Harsh mewing and nasal, raspy spit; males: repetition of a simple phrase.*
NESTING *Deep cup constructed from plant and animal fibers, often incorporating lichens, suspended from twigs by the rim; 3–5 eggs; 1–2 broods; Feb–May.*
FEEDING *Caterpillars, spiders, and flies from leaves usually while perched; berries.*
HABITAT *Mixed evergreen forests year-round; mixed oak-pine woodlands while breeding.*
LENGTH *5in (13cm)*
WINGSPAN *8in (20cm)*

Warbling Vireo

S

Vireo gilvus

This vireo has a cheerful warbling song and a somewhat warbler-like appearance. The eastern subspecies (*V. g. gilvus*), which is heavier and has a larger bill, and the western subspecies (*V. g. swainsonii*) are quite different and may in fact be separate species. Of all the vireos, the Warbling Vireo is most likely to breed in human developments, such as city parks and suburbs.

grayish green upperparts

blackish bill

grayish overall

grayish behind eye

white eyebrow

pale patch between eye and bill

yellowish flanks

pale brownish crown contrasts with darker back

VOICE *Harsh, raspy scold call; males: high, rapid, and highly variable warble.*
NESTING *Rough cup placed high in a deciduous tree, hung from the rim between forked twigs; 3–5 eggs; 2 broods; Mar–Jul.*
FEEDING *Insects, including grasshoppers, aphids, and beetles from leaves; eats fruit in winter.*
HABITAT *Deciduous and mixed forests near water while breeding.*
LENGTH *5½in (14cm)*
WINGSPAN *8½in (21cm)*

Red-eyed Vireo

Vireo olivaceus

Probably the most common songbird of northern and eastern North America, the Red-eyed Vireo is perhaps the quintessential North American vireo, although it is heard more often than it is seen. It sings persistently and monotonously all day long and late into the season, long after other species have stopped singing. It generally stays high in the canopy of the deciduous and mixed woodlands where it breeds. The entire population migrates in the fall to central South America in winter.

head held at downward angle

generally olive above

brown eyes

whitish breast and belly

gray crown

heavy eye-line

long bill

white eyestripe with black upper border

deep red eye

bird appears long and slender

whitish underparts

bluish legs and toes

VOICE *Nasal mewing call; males: slurred three-note phrases, ending in either an upturn or downturn.*
NESTING *Open cup nest of plant fibers bound with spider's web hanging on horizontal fork of tree branch; exterior is sometimes decorated with lichen; 3–5 eggs; 1 brood; May–Jul.*
FEEDING *Insects from leaves, in the canopy and sub-canopy of deciduous trees; primarily fruit during fall and winter.*
HABITAT *Canopy of deciduous forests and pine hardwood forests.*
LENGTH 6in (15cm)
WINGSPAN 10in (25cm)

Philadelphia Vireo

S

Vireo philadelphicus

Despite being widespread, the Philadelphia Vireo remains rather poorly studied. It shares its breeding habitat with the similar looking, but larger and more numerous, Red-eyed Vireo, and, interestingly, it modifies its behavior to avoid competition. It is the most northerly breeding vireo, with its southernmost breeding range barely reaching the US.

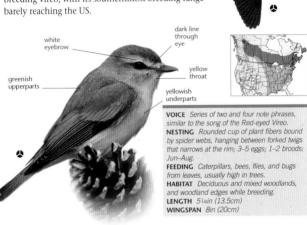

gray cap

slightly hooked, black bill

white eyebrow

dark line through eye

yellow throat

greenish upperparts

yellowish underparts

VOICE *Series of two and four note phrases, similar to the song of the Red-eyed Vireo.*
NESTING *Rounded cup of plant fibers bound by spider webs, hanging between forked twigs that narrows at the rim; 3–5 eggs; 1–2 broods; Jun–Aug.*
FEEDING *Caterpillars, bees, flies, and bugs from leaves, usually high in trees.*
HABITAT *Deciduous and mixed woodlands, and woodland edges while breeding.*
LENGTH *5¼in (13.5cm)*
WINGSPAN *8in (20cm)*

Gray Jay

S

Perisoreus canadensis

Fearless, cunning, and inquisitive, the Gray Jay is adept at stealing food and shiny metal objects, earning it the name of "Camp Robber." In winter, it stores food above the snow for later use by sticking it to trees with its viscous saliva. Gray Jays collect in noisy groups of three to six birds in order to investigate intruders encroaching upon their territory.

long, tail with white corners

dark gray upperparts

white forehead

dark crown

white collar

short bill

gray overall, darker upperparts

brownish back with white streaks

P. c. canadensis **NORTHERN**

black legs and toes

P. c. obscurus **NORTHWESTERN**

dark, smoky gray tail and wings

VOICE *Mostly silent; odd clucks and screeches; sometimes Blue Jay-like jay! and eerie whistles, including bisyllabic whee-oo or ew.*
NESTING *Bulky platform of sticks with cocoons on south side of coniferous tree; 2–5 eggs; 1 brood; Feb–May.*
FEEDING *Insects and berries; also raids birds' nests; raisins, peanuts, bread.*
HABITAT *Coniferous forests, especially lichen-festooned areas with firs and spruces.*
LENGTH *10–11½in (25–29cm)*
WINGSPAN *18in (46cm)*

Blue Jay

S

Cyanocitta cristata

The Blue Jay is common in rural and suburban backyards across Canada and the eastern US. Beautiful as it is, the Blue Jay has a darker side. It often raids the nests of smaller birds for eggs and nestlings. Although usually thought of as a non-migratory species, some Blue Jays undergo impressive migrations, with loose flocks sometimes numbering in the hundreds visible overhead in spring and fall. The Blue Jay is the provincial bird of Prince Edward Island.

long tail with white corners

white streak in blue wings

white patches on wing

white trailing edge feathers

black legs and feet

black patch between eye and bill

blue crest

black collar

plain blue mantle

blue wings and tail

long, black bill

whitish throat

grayish underparts

black bars on tail

VOICE *Harsh, screaming* jay! jay!; *odd ethereal, chortling* queedle-ee-dee; *soft clucks when feeding; can mimic hawk calls.*
NESTING *Cup of strong twigs at variable height in trees or shrubs; 3–6 eggs; 1 brood; Mar–Jul.*
FEEDING *Insects, acorns, small vertebrates, such as lizards, rodents, bird eggs, birds, tree frogs; fruits and seeds.*
HABITAT *Deciduous, coniferous, and mixed woodlands; suburban vegetation and backyards.*
LENGTH *9½–12in (24–30cm)*
WINGSPAN *16in (41cm)*

Steller's Jay

S

Cyanocitta stelleri

Steller's Jay has a blackish breast and mantle, conspicuously crested head, and deep blue body, but plumage varies among local populations. Contrasting head markings are blue in coastal populations, white in the interior, and absent in the Queen Charlotte Islands. Steller's Jay is the provincial bird of British Columbia.

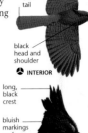

long, blue tail

black head and shoulder

♠ INTERIOR

long, straight bill

white markings on face

slightly darker back

deep blue belly

C. s. macrolopha
INTERIOR

long, black crest

bluish markings on face

dark back and shoulder

blue wings barred with black

black feet and legs

blue belly and breast

C. s. stelleri
PACIFIC

VOICE *Harsh, short, rasping shek, shek, shek; single longer pitch-changing shuhrrrr.*
NESTING *Bulky twig and mud nest, lined with finer plant fibers, close to trunk of tree; 2–6 eggs; 1 brood; Mar–Jun.*
FEEDING *Acorns, pine nuts, fruit; insects and spiders; small vertebrates such as lizards and rodents; raids birds' nests.*
HABITAT *Mainly montane coniferous and mixed forests; also broad-leafed habitats.*
LENGTH *11–12½in (28–32cm)*
WINGSPAN *19in (48cm)*

Black-billed Magpie

S

Pica hudsonia

Loud, flashy, and conspicuous, the Black-billed Magpie is abundant in the northwestern quarter of the continent, confidently strutting across front lawns in some places. Its long tail enables it to make rapid changes in direction in flight. The male will also use his tail to perform a variety of displays while courting a female. Black-billed Magpies sometimes form large flocks in fall.

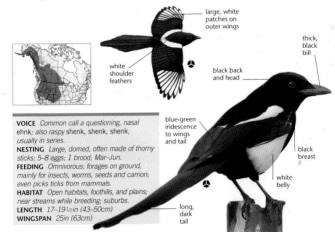

large, white patches on outer wings

white shoulder feathers

thick, black bill

black back and head

blue-green iridescence to wings and tail

black breast

white belly

long, dark tail

VOICE *Common call a questioning, nasal ehnk; also raspy shenk, shenk, shenk, usually in series.*
NESTING *Large, domed, often made of thorny sticks; 5–8 eggs; 1 brood; Mar–Jun.*
FEEDING *Omnivorous; forages on ground, mainly for insects, worms, seeds and carrion; even picks ticks from mammals.*
HABITAT *Open habitats, foothills, and plains; near streams while breeding; suburbs.*
LENGTH *17–19½in (43–50cm)*
WINGSPAN *25in (63cm)*

Clark's Nutcracker

Nucifraga columbiana

A bird of the western mountains, Clark's Nutcracker is an intelligent species that frequents popular scenic overlooks, where it begs for food from visitors. It is named for its dependence on pine nuts, which it forcefully extracts using its powerful feet and chisel-like bill. Nutcrackers cache food when it is abundant—a special throat pouch enables them to carry up to 100 pine nuts per trip.

white patches on black wings

paler face

white undertail feathers

gray body

strong, straight bill

long, black wings

totally white tail from below

black central tail feathers

VOICE Harsh, nasal, rolling rattle, often paired *kraaaa, kraaaa;* also mellower down-slurred *weee-uh,* and frog-like rattle.
NESTING Fine inner cup on stick platform on the side of tree away from trunk; 2–5 eggs; 1 brood; Mar–Jun.
FEEDING Pine nuts; insects, spiders, and carrion; raids nests.
HABITAT Coniferous forests, especially with large-seeded pines.
LENGTH 12in (31cm)
WINGSPAN 24in (62cm)

American Crow

Corvus brachyrhynchos

The American Crow is common in almost all habitats—from wilderness to urban centers. Like most birds with large ranges, there is substantial geographical variation in this species. Birds are black across the whole continent, but size and bill shape vary from region to region. The birds of the coastal Pacific Northwest (*C. b. hesperis*), are on average smaller and have a lower-pitched voice.

black overall

dull black overall

shorter bill

black overall, with greenish sheen

long, black bill

strong legs and feet

VOICE Loud, familiar *caw!;* juveniles' call higher-pitched.
NESTING Stick base with finer inner cup; 3–7 eggs; 1 brood; Apr–Jun.
FEEDING Omnivorous; fruit, carrion, garbage, insects, spiders; raids nests.
HABITAT Prefers open areas with widely spaced, large trees; urban areas, including lawns, parking lots, roadsides, and garbage dumps; partial migrant.
LENGTH 15½–19½in (39–49cm)
WINGSPAN 3ft (1m)

Northwestern Crow Ⓢ

Corvus caurinus

Although smaller, with a lower-pitched voice than the American Crow, the Northwestern Crow is very similar to the American Crow subspecies *C. b. hesperis*. In fact, ornithologists are still debating how closely related the Northwestern and American Crows actually are. The Northwestern Crow is known to dig for clams, pry open barnacles, chase crabs, and catch small fish. It often feeds and roosts in large flocks.

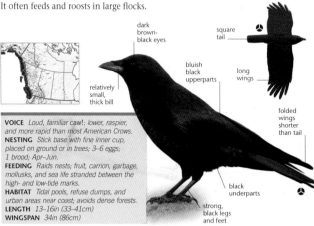

dark brown-black eyes

square tail

bluish black upperparts

long wings

relatively small, thick bill

folded wings shorter than tail

black underparts

strong, black legs and feet

VOICE *Loud, familiar caw!; lower, raspier, and more rapid than most American Crows.*
NESTING *Stick base with fine inner cup, placed on ground or in trees; 3–6 eggs; 1 brood; Apr–Jun.*
FEEDING *Raids nests; fruit, carrion, garbage, mollusks, and sea life stranded between the high- and low-tide marks.*
HABITAT *Tidal pools, refuse dumps, and urban areas near coast; avoids dense forests.*
LENGTH *13–16in (33–41cm)*
WINGSPAN *34in (86cm)*

Common Raven Ⓢ

Corvus corax

The Common Raven is twice the size of the American Crow, a bird of legend and literature. Its Latin name, *Corvus corax*, means "crow of crows." Ravens are perhaps the most intelligent of all birds: they learn quickly, adapt to new circumstances, and communicate with each other through an array of vocal and motional behaviors. The Common Raven is the official bird of the Yukon Territory.

long, black wings

large, protruding head

flared outer wing feathers

thick, long bill, with pronounced curvature

wedge-shaped tail

dark gray neck and underparts

black upperparts, with purplish gloss

shaggy throat

long, black legs and feet

VOICE *Hoarse, rolling krruuk, twangy peals, guttural clicks, and resonant bonks.*
NESTING *Platform of sticks with fine inner material on trees, cliffs, or man-made structure; 4–5 eggs; 1 brood; Mar–Jun.*
FEEDING *Carrion, small crustaceans, fish, rodents, fruit, grain, and garbage; raids nests.*
HABITAT *Tundra, mountainous areas, forests, woodlands, prairies, arid regions, coasts, and near human settlements.*
LENGTH *23½–27in (60–69cm)*
WINGSPAN *4½ft (1.4m)*

Chickadees & Titmice

Chickadees are frequent visitors to backyards and are readily identified by their smooth-looking, dark caps and black bibs. The name "chickadee" is derived from the common calls of several species. Highly social outside the breeding season and generally tolerant of people, they form sociable flocks in winter. Titmice are distinguished from chickadees by their crests; most also have plain throats. Like chickadees, titmice are highly territorial and insectivorous during the breeding season, and then become gregarious seed-eaters afterwards.

TAME BIRDS
Black-capped Chickadees have distinctive black-and-white markings and are often very tame.

Swallows

Swallows are nearly everywhere, except in the polar regions and some of the largest deserts. Ornithologists usually call the short-tailed species martins and the long-tailed ones swallows. All North American swallows are migratory, and most of them winter in Central and South America. They are all superb fliers, and skilled at aerial pursuit and capture of flying insects.

SURFACE SKIMMER
This Tree Swallow flies low over fresh water to catch insects as they emerge into the air.

Cedar Waxwing

S

Bombycilla cedrorum

Flocks of nomadic Cedar Waxwings move around Canada and the US looking for berries, which are their main source of food. Common in a specific location one year, they may disappear the next and occur elsewhere. Northern breeders tend to be more migratory than southern ones. In winter, they may travel as far south as South America. They can often be heard and identified by their calls, long before the flock settles to feed.

black mask

short, yellow tip to tail

streaks on underparts

lacks red on wing

white bars on face

wispy crest

black "bandit" mask

brownish tan back

brown neck and breast

yellow belly

waxy, red tips on inner wing

whitish undertail feathers

VOICE *Shrill trill:* shr-r-r-r-r *or* tre-e-e-e-e-e.
NESTING *Open cup placed in fork of tree, often lined with grasses, plant fibers; 3–5 eggs; 1–2 broods; Jun–Aug.*
FEEDING *Eats in flocks at trees and shrubs with ripe berries throughout the year; also flying insects in summer.*
HABITAT *Woodlands near streams and clearings while breeding; habitats with trees, shrubs, and fruit in winter; spends a lot of time in treetops, but may come down to shrub level.*
LENGTH *7½in (19cm)*
WINGSPAN *12in (30cm)*

Bohemian Waxwing ⓢ

Bombycilla garrulus

The Bohemian Waxwing is the wilder and rarer of the two waxwing species in North America. It breeds mainly in Alaska and western Canada. The species is migratory, but the extent of its wintertime movement is notoriously variable, depending on the availability of wild fruits. They are slightly larger than Cedar Waxwings with distinctive rusty undertail feathers.

wispy crest

gray upperparts

black throat

♀

yellow tail band

variable crest

gray-brown upperparts

♂

yellow edges to outer flight feathers

gray underparts

chestnut undertail feathers

ornate wing markings

VOICE *Dull trill; flocks vocalize constantly with remarkable effect.*
NESTING *Dishevelled cup of sticks and grasses, placed in tree; 4–6 eggs; number of broods unknown; Jun–Jul.*
FEEDING *Insects on the wing in summer; berries of native and exotic trees and shrubs.*
HABITAT *Coniferous forests near disturbed areas while breeding; forest edges, hedges, and residential areas in winter.*
LENGTH *8½in (21cm)*
WINGSPAN *14½in (37cm)*

Mountain Chickadee ⓢ

Poecile gambeli

The Mountain Chickadee is found at elevations of up to 12,000ft (3,600m). Like other chickadees, it stores pine and spruce seeds for harsh winters. Social groups defend their winter territories and food resources, migrating to lower elevations when seeds are scarce. Birds in the Rocky Mountains have a conspicuous white eyebrow and buff-tinged flanks.

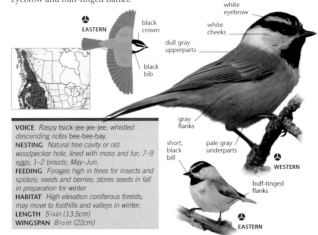

EASTERN

black crown

black bib

white eyebrow

white cheeks

dull gray upperparts

gray flanks

short, black bill

pale gray underparts

WESTERN

buff-tinged flanks

EASTERN

VOICE *Raspy tsick-jee-jee-jee; whistled descending notes bee-bee-bay.*
NESTING *Natural tree cavity or old woodpecker hole, lined with moss and fur; 7–9 eggs; 1–2 broods; May–Jun.*
FEEDING *Forages high in trees for insects and spiders; seeds and berries; stores seeds in fall in preparation for winter.*
HABITAT *High elevation coniferous forests; may move to foothills and valleys in winter.*
LENGTH *5¼in (13.5cm)*
WINGSPAN *8½in (22cm)*

Black-capped Chickadee

Poecile atricapillus

The Black-capped Chickadee is the most widespread chickadee in North America, equally at home in the cold far north and in warm valleys in the south. To cope with the harsh northern winters, this species can decrease its body temperature, entering a controlled hypothermia to conserve energy. Appearance varies according to location, with northern birds being slightly larger and having brighter white wing edgings than southern birds. Although it is a non-migratory species, flocks occasionally travel south of their traditional range in winter. The Black-capped Chickadee is the provincial bird of New Brunswick.

black bib with faded lower margin

white on wings and tail

black-and-white head

grayish brown upperparts

buff flanks fading to white on belly

bright white cheeks

short black bill

white edges on wing feathers

white edges on outer tail feathers

black cap and bib

faded buff flanks

VOICE *Raspy* tsick-a-dee-dee-dee; *loud, clear whistle* bee-bee *or* bee-bee-be, *first note higher in pitch.*
NESTING *Cavity in rotting tree stump, lined with hair, fur, feathers, plant fibers; 6–8 eggs; 1 brood; Apr–Jun.*
FEEDING *Insects and their eggs, and spiders in trees and bushes; mainly seeds in winter; may take seeds from an outstretched hand.*
HABITAT *Wooded habitats including forests, woodlands, parks, and suburbs.*
LENGTH 5¼in (13.5cm)
WINGSPAN 8½in (22cm)

Boreal Chickadee

S

Poecile hudsonicus

The Boreal Chickadee was previously known by other names, including Hudsonian Chickadee and Brown-capped Chickadee. In the past, this species journeyed south of its usual range during winters of food shortage, but this pattern has not occurred in recent decades. Its back color shows geographic variation—grayish in the West and brown in the central and eastern portions of its range.

gray tail

gray cheeks

grayish brown back

brown cap

gray wings

black bib

rich brown flanks and belly

VOICE *Low-pitched, buzzy, lazy* tsee-day-day; *also high-pitched trill,* dididididididi.
NESTING *Cavity lined with fur, hair, plant down; in natural, excavated, or old woodpecker hole; 4–9 eggs; 1 brood; May–Jun.*
FEEDING *Insects, conifer seeds; hoards larvae and seeds in bark crevices in fall in preparation for winter.*
HABITAT *Spruce-fir forests.*
LENGTH *5½in (14cm)*
WINGSPAN *8½in (21cm)*

Chestnut-backed Chickadee

S

Poecile rufescens

The Chestnut-backed is the smallest of all chickadees and possesses the shortest tail. Northern populations have the most brightly colored sides and flanks of all North American chickadees—rich chestnut or rufous, matching the bright back and rump. The Chestnut-backed Chickadee may nest in loose colonies, unlike any other chickadee species.

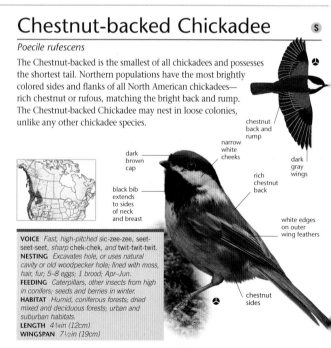

chestnut back and rump

narrow white cheeks

dark brown cap

dark gray wings

black bib extends to sides of neck and breast

rich chestnut back

white edges on outer wing feathers

chestnut sides

VOICE *Fast, high-pitched* sic-zee-zee, seet-seet-seet, *sharp* chek-chek, *and* twit-twit-twit.
NESTING *Excavates hole, or uses natural cavity or old woodpecker hole; lined with moss, hair, fur; 5–8 eggs; 1 brood; Apr–Jun.*
FEEDING *Caterpillars, other insects from high in conifers; seeds and berries in winter.*
HABITAT *Humid, coniferous forests; dried mixed and deciduous forests; urban and suburban habitats.*
LENGTH *4¾in (12cm)*
WINGSPAN *7½in (19cm)*

Tufted Titmouse (S)

Baeolophus bicolor

The Tufted Titmouse is the most widespread of the North American titmice, and one of the two largest and most fearless; it has adapted very well to human habitations. In the last century, its range has expanded significantly northward into southern Canada, probably due to the increased numbers of birdfeeders, which allow it to survive the cold northern winters.

crest may be flattened

gray wings

gray tail

tufted dark gray head

conspicuous black eye in whitish face

prominent orange flanks

black fore-head

gray underparts

gray-black legs and feet

VOICE *Loud, harsh pshurr, pshurr, pshurr; ringing, far-carrying peto peto peto, or peer peer peer.*
NESTING *Tree cavities, old woodpecker holes, and nest boxes, lined with damp leaves, moss, grass, hair; 5–6 eggs; 1 brood; Mar–May.*
FEEDING *Insects, spiders, and their eggs; in winter, corn kernels, seeds, and small fruits.*
HABITAT *Deciduous and coniferous woodlands; parks and gardens.*
LENGTH *6½in (16cm)*
WINGSPAN *10in (25cm)*

Bank Swallow (T)

Riparia riparia

The Bank Swallow is the smallest of North American swallows. It nests in the banks and bluffs of rivers, streams, and lakes, and favors sand and gravel quarries in the East. It breeds from south of the tundra–taiga line down to the central US. The nesting colonies can range from 10 to 2,000 pairs, which are noisy when all the birds are calling.

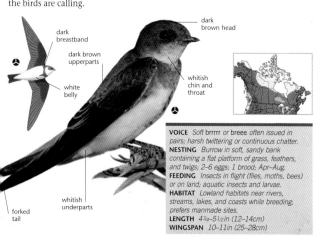

dark breastband

dark brown upperparts

white belly

forked tail

whitish underparts

dark brown head

whitish chin and throat

VOICE *Soft brrrrr or breee often issued in pairs; harsh twittering or continuous chatter.*
NESTING *Burrow in soft, sandy bank containing a flat platform of grass, feathers, and twigs; 2–6 eggs; 1 brood; Apr–Aug.*
FEEDING *Insects in flight (flies, moths, bees) or on land; aquatic insects and larvae.*
HABITAT *Lowland habitats near rivers, streams, lakes, and coasts while breeding; prefers manmade sites.*
LENGTH *4¾–5½in (12–14cm)*
WINGSPAN *10–11in (25–28cm)*

Tree Swallow

Ⓢ

Tachycineta bicolor

One of the most common North American swallows, the Tree Swallow is found from coast to coast in the upper half of the continent all the way up to Alaska. It has iridescent bluish green upperparts and white underparts. Juveniles can be confused with the smaller Bank Swallow, which has a more complete breastband. The Tree Swallow lives in a variety of habitats, but its hole-nesting habit makes it completely dependent on abandoned woodpecker cavities in dead trees and nestboxes.

no blue on head or upperparts

partial grayish brown breastband

dark, pointed wings ♂

slightly forked tail

clean white underparts ♀

small, black bill

iridescent bluish green upperparts

white throat

♂

blackish flight feathers

brilliant white underparts

reddish brown legs and feet

VOICE *Variable high, chirping notes; chatters and soft trills; also complex high and clear two-note whistle phrases.*
NESTING *Layer of fine plant matter in abandoned woodpecker hole or nest box, lined with feathers; 4–6 eggs; 1 brood; May–Jul.*
FEEDING *Swoops after flying insects from dawn to dusk; also forages for bayberries.*
HABITAT *Open habitat close to water such as fields, marshes, and lakes, with standing dead wood while breeding.*
LENGTH *5–6in (13–15cm)*
WINGSPAN *12–14in (30–35cm)*

Violet-green Swallow

Tachycineta thalassina

Although common in its range, the Violet-green Swallow is arguably the least well known of all North American swallows. It often occurs in mountainous conifer forests where it breeds in woodpecker holes in dead trees or in cliff crevices, but it will also use birdhouses. In its mountain habitat, the Violet-green Swallow can be encountered together with the White-throated Swift.

long wings

white underparts

brown on head and sides of face

green head

faint purplish neck patch

emerald green back

purple rump and tail

white, crescent-like patch on throat and face

wing tips extend beyond tail

♀

♂

VOICE *Twittering* chee-chee *of brief duration; dry, brief* zwrack *alarm call.*
NESTING *Nest of grass, twigs, straw, and feathers in a natural or woodpecker cavity in a tree, also cliff or nesting box; 4–6 eggs; 2 broods; Mar–Aug.*
FEEDING *Flying insects (bees, wasps, moths), usually at higher levels than other swallows.*
HABITAT *Open deciduous, coniferous, and mixed woodlands while breeding.*
LENGTH *5in (13cm)*
WINGSPAN *11in (28cm)*

Purple Martin

Progne subis

The Purple Martin is the largest of all North American swallows. Found mostly in the eastern half of the continent, with local populations scattered across the West, this glossy-blue swallow is common in some areas and yet quite scarce in others. In the West it nests in abandoned woodpecker holes, but in the East the Purple Martin now depends almost entirely on "apartment-type" birdhouses for breeding.

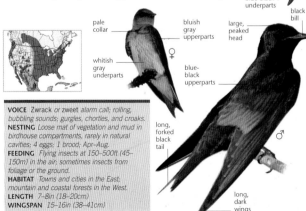

♂

blue-black underparts

black bill

pale collar

bluish gray upperparts

large, peaked head

whitish gray underparts

blue-black upperparts

long, forked black tail

long, dark wings

♀

♂

VOICE *Zwrack or* zweet *alarm call; rolling, bubbling sounds; gurgles, chortles, and croaks.*
NESTING *Loose mat of vegetation and mud in birdhouse compartments, rarely in natural cavities; 4 eggs; 1 brood; Apr–Aug.*
FEEDING *Flying insects at 150–500ft (45–150m) in the air; sometimes insects from foliage or the ground.*
HABITAT *Towns and cities in the East; mountain and coastal forests in the West.*
LENGTH *7–8in (18–20cm)*
WINGSPAN *15–16in (38–41cm)*

Northern Rough-winged Swallow

S

Stelgidopteryx serripennis

The Northern Rough-winged Swallow is found across southern Canada and throughout the US and can be spotted hunting insects over water. In size and habit, it shares many similarities with the Bank Swallow, including breeding habits and color, but the latter's notched tail and smaller size makes it easy to tell them apart.

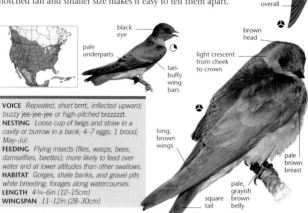

dark face

dark brown overall

black eye

pale underparts

tan-buffy wing bars

brown head

light crescent from cheek to crown

long, brown wings

pale brown breast

pale, grayish brown belly

square tail

VOICE *Repeated, short brrrt, inflected upward; buzzy jee-jee-jee or high-pitched brzzzzzt.*
NESTING *Loose cup of twigs and straw in a cavity or burrow in a bank; 4–7 eggs; 1 brood; May–Jul.*
FEEDING *Flying insects (flies, wasps, bees, damselflies, beetles); more likely to feed over water and at lower altitudes than other swallows.*
HABITAT *Gorges, shale banks, and gravel pits while breeding; forages along watercourses.*
LENGTH *4¾–6in (12–15cm)*
WINGSPAN *11–12in (28–30cm)*

Cliff Swallow

S

Petrochelidon pyrrhonota

The Cliff Swallow is social, sometimes nesting in colonies of over 3,500 pairs, especially in the western US. It is more locally distributed across the east. The adult can best be distinguished by its square tail, orange rump, and white forehead. It is noteworthy for affixing its mud nests to the sides of highway culverts, bridges, and buildings.

long, roundish wings

whitish forehead

dark throat

rusty cheek patch

mottled throat

bluish black cap

rusty-brown cheeks

brown-tinged, black back

pale hind neck collar

bluish black back

pale underparts

pale reddish rump

slight notch in squared tail

VOICE *Purr and churr alarm calls; low, squeaky twitter given in flight and near nests.*
NESTING *Domed nests of mud pellets on cave walls, buildings, culverts, bridges, and dams; 3–5 eggs; 1–2 broods; Apr–Aug.*
FEEDING *Flying insects (often swarming varieties); sometimes forages on the ground; ingests grit to aid digestion.*
HABITAT *Most habitats except deserts, tundra, and unbroken forests.*
LENGTH *5in (13cm)*
WINGSPAN *11–12in (28–30cm)*

Barn Swallow

Hirundo rustica

The Barn Swallow is found just about everywhere in North America south of the Arctic timberline. Originally a cave-nester, it now mostly nests under the eaves of houses, under bridges, and inside buildings such as barns. Steely blue upperparts, reddish underparts, and a deeply forked tail readily identify this bird. While still a fairly common sight, populations of Barn Swallows are in decline across Canada, likely due to loss of nesting sites and foraging habitats.

long, pointed wings

reddish orange underparts

shiny blue head and upperparts

chestnut forehead

duller plumage than adult

deep, chestnut -brown throat

reddish orange belly

slender wings

deeply forked tail

long tail "streamers"

VOICE *High-pitched, squeaky* chee-chee; *series of chatty, pleasant churrs, squeaks, chitterings, and buzzes.*
NESTING *Deep cup of mud and grass-stems attached to vertical surfaces or on ledges; 4–6 eggs; 1–2 broods; May–Sep.*
FEEDING *Flying insects (flies, mosquitoes, wasps, beetles) at lower altitudes than other swallows; sometimes wild berries and seeds.*
HABITAT *Most habitats except tundra; prefers agricultural regions, towns, and highway overpasses.*
LENGTH *6–7½in (15–19cm)*
WINGSPAN *11½–13in (29–33cm)*

Bushtit

Psaltriparus minimus

For much of the year, the Bushtit roams the foothills and valleys of southwestern British Columbia and the western US, in flocks ranging from a handful to many hundreds of birds. Even during the breeding season, the Bushtit retains something of its social nature—raising the young is often a communal affair, with both siblings and single adults helping in the rearing of a pair's chicks.

gray upperparts

mouse-gray upperparts

dark cap

black eye

long tail

tiny, black bill

♂
PACIFIC; BREEDING

pale underparts

black legs and feet

brownish cap

yellow eye

♀
PACIFIC; BREEDING

VOICE *2–3-part soft lisp, ps psss pit, interspersed with hard spit and spick notes.*
NESTING *Enormous pendant structure of cobwebs and leaves, hung from branch; 4–10 eggs; 2 broods; Apr–Jul.*
FEEDING *Spiders and insects from vegetation; often hangs upside-down while feeding.*
HABITAT *Open woodlands and shrubby areas; mainly hillsides in summer; cities and gardens for coastal populations.*
LENGTH *4½in (11.5cm)*
WINGSPAN *6in (15.5cm)*

Horned Lark

Eremophila alpestris

The Horned Lark favors open country, especially places with extensive bare ground. It is characteristic of arid, alpine, and Arctic regions; in these areas, it flourishes in the bleakest of habitats, from sun-scorched, arid lakes in the Great Basin to windswept tundra above the timberline. In some places, the only breeding bird species are the Horned Lark and the Common Raven.

brown wings

tiny "horns"

bold black-and-yellow face

dark streaks on reddish brown upperparts

black tail with narrow, white edges to outer feathers

white underparts

BREEDING

short legs

streaked upperparts

VOICE *Sharp sweet or soo-weet in flight; pleasant, musical tinkling series, followed by sweet... swit... sweet... s'sweea'weea'witta'swit.*
NESTING *In depression in bare ground, sheltered by grass or low shrubs, lined with plant matter; 2–5 eggs; 1–3 broods; Mar–Jul.*
FEEDING *Seeds of grasses and sedges in winter; mostly insects in summer.*
HABITAT *Open habitat with bare ground while breeding, such as short-grass prairies, deserts.*
LENGTH *7in (18cm)*
WINGSPAN *12in (30cm)*

Golden-crowned Kinglet ⓢ

Regulus satrapa

This hardy little bird, barely more than a ball of feathers, breeds in northern and mountainous coniferous forests in Canada and the US, and spends winters across the US and down into the mountain forests of Mexico and Guatemala. Planting of spruce trees in parts of the US Midwest has allowed this species to increase its breeding range in recent years.

whitish wing bars ♂

orange-and-yellow patch on crown, with black border

broad whitish stripe above eye

olive-green upperparts

short, straight bill

white wing bar

notched tail

pale buff to whitish underparts

yellow crown patch, with black border

♂

♀

VOICE *High-pitched* tsee *or see* see; tsee-tsee-tsee-tsee-teet-leetle, *followed by brief trill.*
NESTING *Cup-shaped nest made of moss and bark; 8–9 eggs; 1–2 broods; May–Aug.*
FEEDING *Flies, beetles, mites, spiders, and their eggs from tips of branches, under bark, conifer needles; seeds and persimmon fruits.*
HABITAT *Spruce or fir forests, mixed coniferous–deciduous forests, single-species stands while breeding.*
LENGTH *3¼–4¼in (8–11cm)*
WINGSPAN *5½–7in (14–18cm)*

Ruby-crowned Kinglet ⓢ

Regulus calendula

The Ruby-crowned Kinglet is recognizable because of its very small size, white eye-ring, two white wing bars, and habit of incessantly flicking its wings while foraging. It has a loud, complex song and lays up to 12 eggs in a clutch—probably the highest of any North American songbird. It will sometimes join mixed-species flocks in winter with nuthatches and titmice.

notched tail

patch on crown often concealed

white wing bars

olive-green upperparts

red patch on crown

incomplete white eye-ring

♂

olive underparts

no red patch on crown

♀

VOICE *Husky* jidit; 2–3 clear tee *or* zee, 5–6 lower tu *or* turr, *and repeated* tee-da-leet.
NESTING *Globular nest with enclosed or open cup, made of mosses, feathers, spider's silk, bark, fur; 5–12 eggs; 1 brood; May–Oct.*
FEEDING *Insects, spiders, and their eggs from leaves; fruit and seeds.*
HABITAT *Near water in Black Spruce and tamarack forests, muskegs, mixed conifer and hardwood forests while breeding.*
LENGTH *3½–4¼in (9–11cm)*
WINGSPAN *6–7in (15–18cm)*

Wrens

Generally dull-colored, most wrens are shades of brown with light and dark streaking, and some have furtive habits. Wrens are also renowned in the avian world for their remarkable songs, and, in some species, for singing precisely synchronized duets.

COCKED TAIL
As they sing, Winter Wrens often hold their tails upward in a near-vertical position.

Nuthatches

These common, plump-bodied woodland birds use their straight, pointed bills to probe for insects and spiders in tree crevices. Referred to as "topsy-turvy birds," their strong feet and long claws allow them to move downwards and upside-down along the underside of branches in search of food.

ACROBATIC POSE
Downward-facing nuthatches such as this one often lift their heads in a characteristic pose.

Thrashers

Thrashers are known for their ability to mimic the songs of other species in their own song sequences. They are characterized by their long, curved bills and somewhat reclusive habits, though some, like the Northern Mockingbird, are brash, conspicuous, and sport a short, somewhat straight bill.

DISTINCTIVE BILL
This Brown Thrasher is characterized by its slender, curved bill, long, thin legs, and long, rounded tail.

Rock Wren

Ⓢ

Salpinctes obsoletus

An inhabitant of rocky landscapes, the Rock Wren runs, flutters, and darts in and out of crevices in search of food, and is known to bob and sway conspicuously when a human approaches. Its oddest habit is to "pave" the area in front of its nest entrance with a walkway of pebbles. Its voice, while not particularly loud, carries surprisingly far through the dry air of the West.

rusty rump

faint eyestripe

grayish brown upperparts

bright buff feather tips

black-and-white barring on undertail

thin, brown streaks

pale buff lower flanks

pale lemon-yellow belly

VOICE Sharp ch'keer; repeated phrases such as chuwee chuwee, teedee teedee, reminiscent of a mockingbird or a thrasher.
NESTING Cup of various grasses lined with soft materials, in rock crevice or under overhang; 4–8 eggs; 1–2 broods; Apr–Aug.
FEEDING Probes in rock crevices on ground and in dirt banks for insects and spiders.
HABITAT Arid country with rocky cliffs and canyons; manmade quarries and gravel piles.
LENGTH 6in (15cm)
WINGSPAN 9in (23cm)

Canyon Wren

Ⓢ

Catherpes mexicanus

The Canyon Wren's loud, clear whistles echo across canyons in the West, but the singer usually stays out of sight, remaining high among the crevices of its cliffside home. It is known for its extraordinary ability to walk up, down, and sideways on vertical rock walls; its strong toes and long claws can find a grip in the tiniest of depressions in the rock.

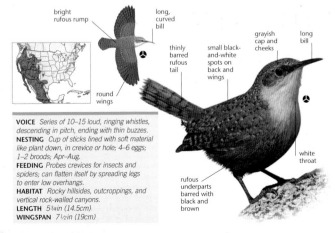

bright rufous rump

long, curved bill

thinly barred rufous tail

small black-and-white spots on back and wings

grayish cap and cheeks

long bill

round wings

white throat

rufous underparts barred with black and brown

VOICE Series of 10–15 loud, ringing whistles, descending in pitch, ending with thin buzzes.
NESTING Cup of sticks lined with soft material like plant down, in crevice or hole; 4–6 eggs; 1–2 broods; Apr–Aug.
FEEDING Probes crevices for insects and spiders; can flatten itself by spreading legs to enter low overhangs.
HABITAT Rocky hillsides, outcroppings, and vertical rock-walled canyons.
LENGTH 5¾in (14.5cm)
WINGSPAN 7½in (19cm)

Sedge Wren

S

Cistothorus platensis

A shy bird, the Sedge Wren stays out of sight except when singing atop a sedge stalk or shrub. If discovered, it flies a short distance, drops down, and runs to hide in vegetation. The Sedge Wren breeds from May–June in the north central region of its range, and July–September in the southern and eastern regions. The male will build up to 8–10 unlined "dummy" nests before the female builds the better-concealed, real nest.

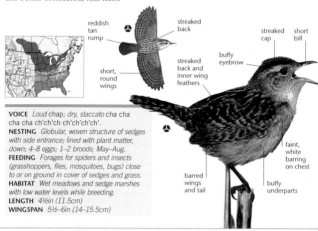

reddish tan rump

streaked back

streaked cap

short bill

buffy eyebrow

streaked back and inner wing feathers

short, round wings

faint, white barring on chest

barred wings and tail

buffy underparts

VOICE *Loud chap; dry, staccato cha cha cha cha ch'ch'ch ch'ch'ch'ch'.*
NESTING *Globular, woven structure of sedges with side entrance; lined with plant matter, down; 4–8 eggs; 1–2 broods; May–Aug.*
FEEDING *Forages for spiders and insects (grasshoppers, flies, mosquitoes, bugs) close to or on ground in cover of sedges and grass.*
HABITAT *Wet meadows and sedge marshes with low water levels while breeding.*
LENGTH *4½in (11.5cm)*
WINGSPAN *5½–6in (14–15.5cm)*

Marsh Wren

S

Cistothorus palustris

The Marsh Wren, a common resident of saltwater and freshwater marshes, is known for singing loudly through both day and night. The males perform aerial courtship flights while singing, and mate with two or more females. The male builds several dummy nests before his mate constructs one herself. Eastern (*C. p. palustris*) and Western (*C. p. paludicola*) Marsh Wrens differ in voice and behavior, and are thought by some to be separate species.

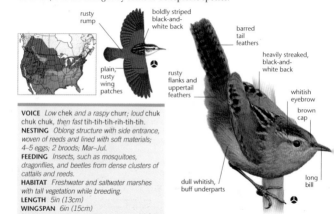

rusty rump

boldly striped black-and-white back

barred tail feathers

heavily streaked, black-and-white back

plain, rusty wing patches

rusty flanks and uppertail feathers

whitish eyebrow

brown cap

VOICE *Low chek and a raspy churr; loud chuk chuk chuk, then fast tih-tih-tih-rih-tih-tih.*
NESTING *Oblong structure with side entrance, woven of reeds and lined with soft materials; 4–5 eggs; 2 broods; Mar–Jul.*
FEEDING *Insects, such as mosquitoes, dragonflies, and beetles from dense clusters of cattails and reeds.*
HABITAT *Freshwater and saltwater marshes with tall vegetation while breeding.*
LENGTH *5in (13cm)*
WINGSPAN *6in (15cm)*

dull whitish, buff underparts

long bill

Bewick's Wren

Thryomanes bewickii

Bewick's Wren is familiar around human habitations, nesting in any sort of hole or crevice in barns, houses, abandoned machinery, woodpiles, and even trash heaps. This wren has undergone large-scale changes in its geographic distribution: in the 19th century its range expanded northward to the eastern and midwestern US, but it gradually disappeared from those regions in the 20th century.

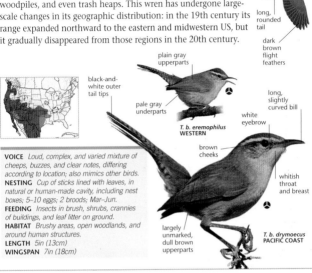

long, rounded tail

dark brown flight feathers

plain gray upperparts

black-and-white outer tail tips

pale gray underparts

T. b. eremophilus
WESTERN

long, slightly curved bill

white eyebrow

brown cheeks

whitish throat and breast

largely unmarked, dull brown upperparts

T. b. drymoecus
PACIFIC COAST

VOICE Loud, complex, and varied mixture of cheeps, buzzes, and clear notes, differing according to location; also mimics other birds.
NESTING Cup of sticks lined with leaves, in natural or human-made cavity, including nest boxes; 5–10 eggs; 2 broods; Mar–Jun.
FEEDING Insects in brush, shrubs, crannies of buildings, and leaf litter on ground.
HABITAT Brushy areas, open woodlands, and around human structures.
LENGTH 5in (13cm)
WINGSPAN 7in (18cm)

Carolina Wren

Thryothorus ludovicianus

The Carolina Wren is rarely still, often flicking its tail and looking around nervously. Extremely harsh winters at the northernmost fringe of the Carolina Wren's range in southeastern Canada, to where it has expanded, can cause a sudden decline in numbers as food resources are covered for long periods by ice and heavy snow. At such times, survival may depend on human help for food and shelter.

thin, black barring on tail

white wing spots

huge head

duller overall

tiny tail

FLEDGLING

white eyebrow bordered by black above

powerful-looking, bluish bill

white spots on wing

rufous upperparts

pinkish legs and toes

buffy underparts

VOICE Sharp chlip or long, harsh chatter; loud, long, fast whee'dle-dee whee'dle-dee whee'dle-dee.
NESTING Cup of weeds, twigs, leaves in natural or human-made cavity; 4–8 eggs; 2–3 broods; Apr–Jul.
FEEDING Insects in shrubs and on ground; in winter, prefers peanut butter or suet at a feeder.
HABITAT Bushy woodland habitats such as thickets, parks, and gardens while breeding.
LENGTH 5¼in (13.5cm)
WINGSPAN 7½in (19cm)

House Wren

S

Troglodytes aedon

Of all the North American wrens, the House Wren is the plainest, yet one of the most familiar and endearing. However, it can be a fairly aggressive species, driving away nearby nesting birds of its own species and others by destroying nests, puncturing eggs, and even killing young. In the 1920s, distraught bird lovers campaigned to eradicate the House Wren, though the campaign did not last long as most were fortunately in favor of letting nature take its course.

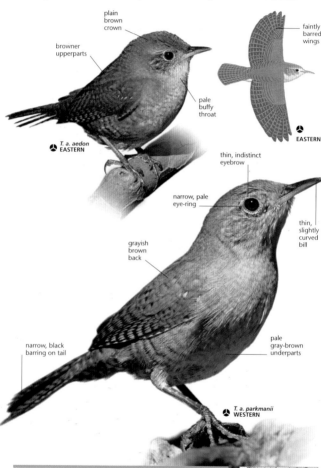

plain brown crown

browner upperparts

faintly barred wings

pale buffy throat

T. a. aedon EASTERN

EASTERN

thin, indistinct eyebrow

narrow, pale eye-ring

thin, slightly curved bill

grayish brown back

narrow, black barring on tail

pale gray-brown underparts

T. a. parkmanii WESTERN

VOICE *Sharp chep or cherr; several short notes, followed by bubbly explosion of spluttering notes.*
NESTING *Cup lined with soft material on stick platform in natural, manmade cavities, such as nest boxes; 5–8 eggs; 2–3 broods; Apr–Jul.*
FEEDING *Insects and spiders from trees and shrubs, gardens, and yards.*
HABITAT *Cities, towns, parks, farms, yards, gardens, and woodland edges while breeding.*
LENGTH *4½in (11.5cm)*
WINGSPAN *6in (15cm)*

Winter Wren ⓢ

Troglodytes troglodytes

The Winter Wren has one of the loudest songs of any North American bird of a similar size: the male's song carries far through its forest haunts. It is a widespread breeder, found from the Aleutians and Alaska eastward to Newfoundland, and as far south as California in the West and the Appalachians in the East, where the subspecies *T. t. pullus* resides.

short, barred tail

barred, rounded wings

stubby tail, usually cocked straight up

dark brown, barred back

distinct, tan eyebrow

small, thin bill

flanks strongly barred

VOICE Double chek-chek or chimp-chimp; complex series of warbles, trills, single notes.
NESTING Messy mound lined with feathers in well-hidden cavity near ground with dead wood and crevices; 4–7 eggs; 1–2 broods; Apr–Jul.
FEEDING Insects from low, dense undergrowth, often along streams; sometimes thrusts its head into water to capture prey.
HABITAT Prefers evergreen trees with dense understory and streams banks while breeding.
LENGTH 4in (10cm)
WINGSPAN 5½in (14cm)

Blue-gray Gnatcatcher ⓢ

Polioptila caerulea

The Blue-gray Gnatcatcher spends much of its time foraging high up in tall trees, giving its continual wheezy call; in winter, it is generally silent. This species is the most northerly of the North American gnatcatchers and is also the only one to migrate. It can exhibit aggressive behavior and is capable of driving off considerably larger birds than itself.

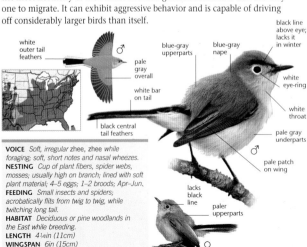

white outer tail feathers

♂

pale gray overall

blue-gray upperparts

blue-gray nape

black line above eye; lacks it in winter

white eye-ring

white bar on tail

white throat

black central tail feathers

pale gray underparts

♂

pale patch on wing

lacks black line

paler upperparts

♀

VOICE Soft, irregular zhee, zhee while foraging; soft, short notes and nasal wheezes.
NESTING Cup of plant fibers, spider webs, mosses; usually high on branch; lined with soft plant material; 4–5 eggs; 1–2 broods; Apr–Jun.
FEEDING Small insects and spiders; acrobatically flits from twig to twig, while twitching long tail.
HABITAT Deciduous or pine woodlands in the East while breeding.
LENGTH 4¼in (11cm)
WINGSPAN 6in (15cm)

Pygmy Nuthatch Ⓢ

Sitta pygmaea

Pygmy Nuthatches are found in noisy and busy flocks throughout the year in their pine forest home of western North America. They are cooperative breeders, with young birds from the previous year's brood often helping adult birds raise the next year's young. They prefer Ponderosa and Jeffery Pines and are often absent from mountain ranges that lack their favorite trees.

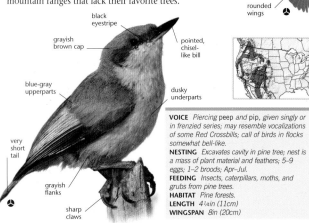

rounded wings

black eyestripe

grayish brown cap

pointed, chisel-like bill

blue-gray upperparts

dusky underparts

very short tail

grayish flanks

sharp claws

VOICE *Piercing* peep *and* pip, *given singly or in frenzied series; may resemble vocalizations of some Red Crossbills; call of birds in flocks somewhat bell-like.*
NESTING *Excavates cavity in pine tree; nest is a mass of plant material and feathers; 5–9 eggs; 1–2 broods; Apr–Jul.*
FEEDING *Insects, caterpillars, moths, and grubs from pine trees.*
HABITAT *Pine forests.*
LENGTH *4¼in (11cm)*
WINGSPAN *8in (20cm)*

White-breasted Nuthatch Ⓢ

Sitta carolinensis

The amiable White-breasted Nuthatch inhabits residential neighborhoods across the US and southern Canada, and often visits birdfeeders in winter. The largest of our nuthatches, it spends more time probing on trunks and boughs than other nuthatches do, walking forward, backward, upside-down, or horizontally. Five subspecies occur in Canada and in the US. They differ in call notes and, to a lesser extent, in plumage.

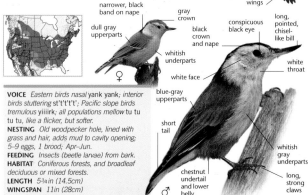

white flashes on tail

♂

rounded wings

narrower, black band on nape

gray crown

dull gray upperparts

black crown and nape

conspicuous black eye

long, pointed, chisel-like bill

whitish underparts

white face

♀

white throat

white face

blue-gray upperparts

whitish gray underparts

short tail

chestnut undertail and lower belly

♂

long, strong claws

VOICE *Eastern birds nasal* yank yank; *interior birds stuttering* st't't't't'; *Pacific slope birds tremulous* yiiiirk; *all populations mellow* tu tu tu tu, *like a flicker, but softer.*
NESTING *Old woodpecker hole, lined with grass and hair, adds mud to cavity opening; 5–9 eggs, 1 brood; Apr–Jun.*
FEEDING *Insects (beetle larvae) from bark.*
HABITAT *Coniferous forests, and broadleaf deciduous or mixed forests.*
LENGTH *5¾in (14.5cm)*
WINGSPAN *11in (28cm)*

Red-breasted Nuthatch

Sitta canadensis

This aggressive, inquisitive nuthatch, with its distinctive black eye stripe, breeds in conifer forests across North America. The bird inhabits mountains in the West; in the East, it is found in lowlands and hills. However, sometimes it breeds in conifer groves away from its core range. Each fall, birds move from their main breeding grounds, but the extent of this exodus varies from year to year, depending on population cycles and food availability.

dark blue-gray crown and eyestripe

slightly muted head pattern

rounded wings

♀

white bands on tail

♂

pale orange underparts

bold black-and-white head pattern

pointed, chisel-like bill

black eyestripe

blue-gray upperparts

white cheeks

♂

rusty underparts

blue-gray, short tail, with black side feathers

compact body shape

VOICE *One-note tooting sound, often repeated, with nasal yet musical quality: aaank, enk, ink, like a horn.*
NESTING *Excavates cavity in pine tree; nest of grass lined with feathers, with sticky pine resin applied to entrance to thwart predators; 5–7 eggs, 1 brood; May–Jul.*
FEEDING *Beetle grubs from bark; also insect larvae found on conifer needles; seeds in winter.*
HABITAT *Coniferous and mixed hardwood forests.*
LENGTH *4¼ in (11cm)*
WINGSPAN *8½ in (22cm)*

Brown Creeper ⓢ

Certhia americana

Although fairly common, the Brown Creeper is one of the most understated of the forest birds, with its soft vocalizations and cryptic plumage. As it forages, it hops up a tree trunk, then flies down to another tree, starts again from near the ground, hops up, and so on. The Brown Creeper is a partial migrant—some move south in the fall; others remain close to their breeding grounds.

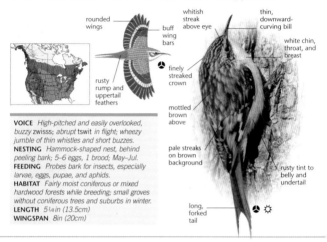

rounded wings

buff wing bars

rusty rump and uppertail feathers

finely streaked crown

whitish streak above eye

thin, downward-curving bill

white chin, throat, and breast

mottled brown above

pale streaks on brown background

rusty tint to belly and undertail

long, forked tail

VOICE High-pitched and easily overlooked, buzzy *zwisss*; abrupt *tswit* in flight; wheezy jumble of thin whistles and short buzzes.
NESTING Hammock-shaped nest, behind peeling bark; 5–6 eggs, 1 brood; May–Jul.
FEEDING Probes bark for insects, especially larvae, eggs, pupae, and aphids.
HABITAT Fairly moist coniferous or mixed hardwood forests while breeding; small groves without coniferous trees and suburbs in winter.
LENGTH 5¼ in (13.5cm)
WINGSPAN 8in (20cm)

Gray Catbird ⓢ

Dumetella carolinensis

In addition to its feline-like, mewing calls, the Gray Catbird can also sing two notes simultaneously. It has been reported to imitate the vocalizations of over 40 bird species, at least one frog species, and several sounds produced by machines and electronic devices. Despite their shy, retiring nature, Gray Catbirds tolerate human presence and will rest in shrubs in suburban and urban lots.

long, black tail

gray overall

gray upperparts

dark gray to black head

straight, blackish bill

large, black eye

bright brick-red undertail feathers

gray underparts

VOICE Mew call, like a young kitten; long, complex series of unhurried notes, sometimes interspersed with whistles and squeaks.
NESTING Large, untidy cup of woven twigs, grass, and hair lined with finer material; 3–4 eggs; 1–2 broods; May–Aug.
FEEDING Wide variety of berries and insects, whatever is most abundant in season.
HABITAT Mixed young to mid-aged forests with abundant undergrowth while breeding.
LENGTH 8–9½ in (20–24cm)
WINGSPAN 10–12in (25–30cm)

Northern Mockingbird

Mimus polyglottos

An impressive mimic, the Northern Mockingbird can incorporate over 100 different phrases of as many different birds in their songs. Its range has expanded in the last few decades, due partly to its high tolerance for human habitats. It frequently gives a "wing-flash" display, showing its white outer wing feather patches when holding its wings overhead, purportedly to startle insects into revealing themselves.

shorter tail

speckled breast and belly

white patches on wing

pointed, curved bill

gray head

yellow eye

long tail with white outer tail feathers

white undertail feathers

white patch on wing feathers

VOICE *Long, complex repertoire often imitating other birds, non-bird noises, and the sounds of mechanical devices.*
NESTING *Bulky cup of twigs, lined, in shrub or tree; 3–5 eggs; 1–3 broods; Mar–Aug.*
FEEDING *Wide variety of fruit, berries, and insects (ants, beetles, and grasshoppers).*
HABITAT *Along edges of disturbed habitats such as young forests and suburban and urban areas with shrubs or hedges.*
LENGTH *8½–10in (22–25cm)*
WINGSPAN *13–15in (33–38cm)*

Sage Thrasher

Oreoscoptes montanus

This plain-colored bird is the smallest of the North American thrashers. It can recognize and remove the eggs of brood parasites, especially those of the Brown-headed Cowbird. Unfortunately, it may also be the least studied of the thrasher group, perhaps because the dense nature of its sagebrush habitat makes study difficult.

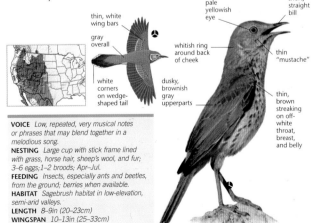

thin, white wing bars

gray overall

white corners on wedge-shaped tail

pale yellowish eye

short, straight bill

whitish ring around back of cheek

dusky, brownish gray upperparts

thin "mustache"

thin, brown streaking on off-white throat, breast, and belly

VOICE *Low, repeated, very musical notes or phrases that may blend together in a melodious song.*
NESTING *Large cup with stick frame lined with grass, horse hair, sheep's wool, and fur; 3–6 eggs; 1–2 broods; Apr–Jul.*
FEEDING *Insects, especially ants and beetles, from the ground; berries when available.*
HABITAT *Sagebrush habitat in low-elevation, semi-arid valleys.*
LENGTH *8–9in (20–23cm)*
WINGSPAN *10–13in (25–33cm)*

Brown Thrasher ⓢ

Toxostoma rufum

The Brown Thrasher, another well-known mimic, usually keeps to dense underbrush and prefers running or hopping to flying. When nesting, it can recognize and remove the eggs of brood parasites like the Brown-headed Cowbird. The Brown Thrasher is an energetic singer and has one of the largest repertoires of any songbird in North America.

rufous wings and upperparts

fairly straight, dark bill

bright yellow eye

grayish cheeks

reddish brown upperparts

long tail with pale outer tips

indistinct "mustache"

two pale wing bars

dark streaking on pale underparts

long tail, paler than back

VOICE Rasping sounds; long series of musical notes, often imitating other species; repeats phrase twice before moving onto the next one.
NESTING Bulky cup of twigs, close to ground, lined with leaves, grass, bark; 3–5 eggs; 1 brood; Apr–Jul.
FEEDING Mainly beetles and worms from leaf litter; cultivated grains, nuts, berries, and fruit.
HABITAT Densely wooded habitats with thick undergrowth; woodland edges, riverside trees.
LENGTH 10–12in (25–30cm)
WINGSPAN 11–14in (28–36cm)

European Starling ⓢ

Sturnus vulgaris

This non-native species is perhaps the most successful bird in North America—and probably the most maligned. In the 1890s, 100 European Starlings were released in New York City's Central Park; these were the ancestors of the millions of birds that now live across Canada and the US. These birds compete with native species like kestrels for nesting cavities, and usually win.

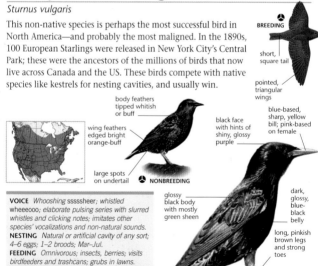

BREEDING

short, square tail

pointed, triangular wings

body feathers tipped whitish or buff

wing feathers edged bright orange-buff

large spots on undertail

NONBREEDING

black face with hints of shiny, glossy purple

blue-based, sharp, yellow bill; pink-based on female

glossy black body with mostly green sheen

dark, glossy, blue-black belly

long, pinkish brown legs and strong toes

♂ BREEDING

VOICE Whooshing ssssheer; whistled wheeeooo; elaborate pulsing series with slurred whistles and clicking notes; imitates other species' vocalizations and non-natural sounds.
NESTING Natural or artificial cavity of any sort; 4–6 eggs; 1–2 broods; Mar–Jul.
FEEDING Omnivorous; insects, berries; visits birdfeeders and trashcans; grubs in lawns.
HABITAT Cities, towns, and farmlands; also relatively "wild" settings far from people.
LENGTH 8½in (21cm)
WINGSPAN 16in (41cm)

Thrushes

Most thrushes are medium-sized brown- or olive-brown-backed birds with varying amounts of spotting or speckling underneath and beautiful flutelike songs. By contrast, the Varied Thrush, with its bold black-and-rust pattern, the brightly colored bluebirds, and the Townsend's Solitaire are all striking enough to stand out from other thrushes.

GROUND BIRD
Though they perch to sing, thrushes, including this Varied Thrush, spend a lot of their time on or near the ground.

Wagtails & Pipits

Wagtails are named for their habit of constantly bobbing their long, slender tails up and down. Although primarily a European genus, one species is considered a regular breeder in Canada, and two others are routinely sighted along the Bering Sea coast and Aleutian Islands.

Unlike wagtails, the two species of pipit that breed in North America also winter there. Very much birds of open, treeless country, both pipit species are likely to be seen on their widespread wintering grounds more often than in their breeding range.

COUNTRY-LOVERS
Pipits, such as this female American Pipit, prefer to live in open countryside.

Varied Thrush

D

Ixoreus naevius

Found in old-growth forests of southern Alaska and British
Columbia, the Varied Thrush is the most beautiful of the
thrushes, with a haunting, ethereal song. It is rather shy and
not always easy to spot, except when bringing food to its
nestlings. The Varied Thrush's orange and black head, deep bluish
black back, and its two rusty wing bars make it easy to identify.

white
double
underwing
bar

♂

black
breastband

brownish
gray
upperparts

orange breast
and throat
with distinct
spotting

orange
eyebrow

black
cheeks

♀

rusty
orange
patches

bluish gray
upperparts

♂

rusty
orange
breast,
faintly
spotted

white
undertail
feathers

VOICE *Single rising or falling note; repeats.*
NESTING *Bulky cup of twigs, dead leaves,
bark, grass, and weeds, lined with fine grass
stems, mainly in conifer trees, against trunk;
3–4 eggs; 1–2 broods; Apr–Aug.*
FEEDING *Insects and caterpillars while
breeding; fruit and berries in winter.*
HABITAT *Moist coniferous forests,
preferably mature, while breeding; ravines,
thickets, and suburban lawns in winter.*
LENGTH *7–10in (18–25cm)*
WINGSPAN *13–15in (33–38cm)*

Western Bluebird

S

Sialia mexicana

Very similar to its close relative, the Eastern Bluebird, the male
Western Bluebird is endowed with a spectacular plumage—
brilliant blue upperparts and deep chestnut-orange underparts.
Unlike the Eastern Bluebird, the Western Bluebird has a brown
back and a complete blue hood. Females and juveniles are
harder to distinguish, but their ranges are quite different.

pale
blue
belly

short
wings
and tail

♂

blue
hood

brownish
back

grayish
throat

♀

rust
patch
on back

blue
shoulder
and wing

♂

blue
wings
and tail

chestnut
breast
and
flanks

pale
blue
belly

rusty
undertail

VOICE *Soft few, few or fewrr-fewrr; pleasant,
soft series of churring notes, all strung together,
often given at dawn.*
NESTING *Shallow cup of dry grass and
feathers in natural tree cavity or old woodpecker
cavity; 4–6 eggs; 1–2 broods; Mar–Jul.*
FEEDING *Mainly insects in breeding season;
berries, such as juniper, in winter.*
HABITAT *Open coniferous and deciduous
woodlands while breeding.*
LENGTH *6–7in (15–18cm)*
WINGSPAN *11½–13in (29–33cm)*

Eastern Bluebird

Sialia sialis

The Eastern Bluebird's vibrant blue and chestnut body is a beloved sight in eastern North America, especially after the remarkable comeback of the species in the past 30 years. After much of the bird's habitat was eliminated by agriculture in the mid-1900s, volunteers offered the bluebirds nest boxes as alternatives to their tree cavities, and they took to these like ducks to water. The Eastern Bluebird's mating system can entail males breeding with multiple partners.

gray-brown upperparts

spotted throat and breast

♂

bluish gray underwings

white belly

rufous breast and throat

pale chestnut throat

♀

gray upperparts

blue wings, rump, and tail

♂

bright blue upperparts

chestnut brown chin, throat, breast, and flanks

white belly

white undertail

VOICE Melodious series of soft, whistled notes; churr-wi or churr-li.
NESTING Cavity nester, in trees or provided boxes; nest of grass lined with grass, weeds, and twigs; uses old nests of other species; 3–7 eggs; 2 broods; Feb–Sep.
FEEDING Insects in breeding season, such as grasshoppers and caterpillars; also fruits and plants in winter.
HABITAT Clearings and woodland edges; open habitats in rural, urban, and suburban areas.
LENGTH 6–8in (15–20cm)
WINGSPAN 10–13in (25–33cm)

Mountain Bluebird ⓢ

Sialia currucoides

A bird of the West, especially sub-alpine meadows, the Mountain Bluebird differs from the other two *Sialia* species by lacking any reddish chestnut in its plumage. It is also more slender and flies in an almost lazy manner. More often than its two relatives, it feeds by hovering, kestrel-like, over meadows, before pouncing on insects. Males guard their mates from pair-bond to egg-hatching time.

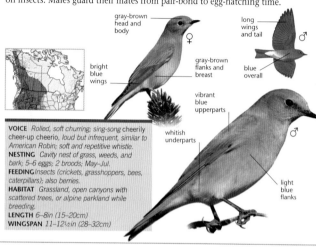

gray-brown head and body

long wings and tail ♂

bright blue wings

gray-brown flanks and breast

blue overall

vibrant blue upperparts

whitish underparts

♂

light blue flanks

VOICE Rolled, soft churring; sing-song cheerily cheer-up cheerio, loud but infrequent, similar to American Robin; soft and repetitive whistle.
NESTING Cavity nest of grass, weeds, and bark; 5–6 eggs; 2 broods; May–Jul.
FEEDING Insects (crickets, grasshoppers, bees, caterpillars); also berries.
HABITAT Grassland, open canyons with scattered trees, or alpine parkland while breeding.
LENGTH 6–8in (15–20cm)
WINGSPAN 11–12½in (28–32cm)

Townsend's Solitaire ⓢ

Myadestes townsendi

The shy Townsend's Solitaire inhabits most of the West, especially high-elevation coniferous forests of the Sierras and Rockies. Its plumage remains the same throughout the year, and the sexes look alike. From a perch high on a branch, Townsend's Solitaire darts after flying insects and audibly snaps its bill shut after catching its prey, unlike other thrush-like birds.

dark gray outer flight feathers

long tail

wide, buff bands on flight feathers

short head

spotted back

heavily spotted breast

long tail

short, black bill

large, black eye

white eye-ring

gray upperparts and head

paler underparts

pale chestnut-tan patches

long, dark tail with white outer feathers

VOICE Single-note, high whistles; sings all year; robin-like trills and squeaky notes.
NESTING Cup of pine needles, dry grass, weed stems, and bark on ground or under overhang; 4 eggs; 1–2 broods; May–Aug.
FEEDING Insects and spiders while breeding; fruits and berries, particularly junipers.
HABITAT Open conifer forests along steep slopes or landslide areas while breeding; open woodlands with junipers in winter.
LENGTH 8–8½in (20–22cm)
WINGSPAN 13–14½in (33–37cm)

Veery

(D)

Catharus fuscescens

The Veery is medium-sized, like the other *Catharus* thrushes, but browner overall and less spotted. There is geographic variation to its plumage; four subspecies have been described to reflect this. Eastern birds (*C. f. fuscescens*) are ruddier than their western relations (*C. f. salicicola*). The Veery migrates to spend the northern winter months in central Brazil.

pale, reddish brown upperparts

brownish tan upperparts

less distinct spotting on breast

creamy pink at base of bill

C. f. fuscescens EASTERN

inconspicuous, pale eye-ring

black upper bill

white underparts

poorly marked brown spots on buff breast and throat

tan wash on flanks

creamy pink legs and feet

C. f. salicicola WESTERN

VOICE *Series of descending da-ve-ur, vee-ur, veer, veer, somewhat bi-tonal, sounding like the name Veery; rather soft veer.*
NESTING *Cup of dead leaves, bark, weed stems, and moss on or near ground; 4 eggs; 1–2 broods; May–Jul.*
FEEDING *Insects, spiders, snails from ground; fruit and berries after breeding.*
HABITAT *Canopy in damp deciduous forests or habitat near rivers in summer.*
LENGTH *7in (18cm)*
WINGSPAN *11–11½in (28–29cm)*

Gray-cheeked Thrush

(S)

Catharus minimus

The Gray-cheeked Thrush is the least known of the four North American *Catharus* thrushes because it breeds in remote areas of Canada and Alaska. During migration, the Gray-cheeked Thrush is more likely to be heard in flight at night than seen on the ground by birdwatchers.

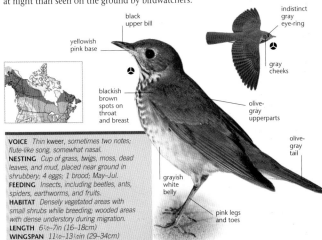

black upper bill

yellowish pink base

indistinct gray eye-ring

gray cheeks

blackish brown spots on throat and breast

olive-gray upperparts

olive-gray tail

grayish white belly

pink legs and toes

VOICE *Thin kweer, sometimes two notes; flute-like song, somewhat nasal.*
NESTING *Cup of grass, twigs, moss, dead leaves, and mud, placed near ground in shrubbery; 4 eggs; 1 brood; May–Jul.*
FEEDING *Insects, including beetles, ants, spiders, earthworms, and fruits.*
HABITAT *Densely vegetated areas with small shrubs while breeding; wooded areas with dense understory during migration.*
LENGTH *6½–7in (16–18cm)*
WINGSPAN *11½–13½in (29–34cm)*

Bicknell's Thrush

Catharus bicknelli

Previously considered a subspecies of the Gray-cheeked Thrush, Bicknell's Thrush is best distinguished by its song, which is less full and lower in pitch than the Gray-cheeked Thrush. It breeds only in dwarf conifer forests on mountain tops in the northeastern US and adjacent Canada, usually above 3,000ft (1,000m). They mate with multiple partners in a single season, and males may care for young in multiple nests.

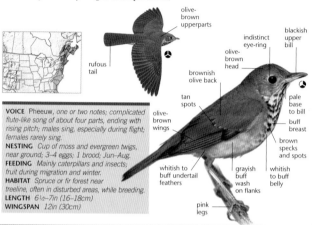

olive-brown upperparts

rufous tail

indistinct eye-ring

blackish upper bill

olive-brown head

brownish olive back

tan spots

olive-brown wings

pale base to bill

buff breast

brown specks and spots

whitish to buff undertail feathers

grayish buff wash on flanks

whitish to buff belly

pink legs

VOICE *Pheeuw, one or two notes; complicated flute-like song of about four parts, ending with rising pitch; males sing, especially during flight; females rarely sing.*
NESTING *Cup of moss and evergreen twigs, near ground; 3–4 eggs; 1 brood; Jun–Aug.*
FEEDING *Mainly caterpillars and insects; fruit during migration and winter.*
HABITAT *Spruce or fir forest near treeline, often in disturbed areas, while breeding.*
LENGTH *6½–7in (16–18cm)*
WINGSPAN *12in (30cm)*

Swainson's Thrush

Catharus ustulatus

Swainson's Thrush can be distinguished from other spotted thrushes by its buffy face and the rising pitch of its flute-like, melodious song. It also feeds higher up in the understory than most of its close relatives. The western subspecies of Swainson's Thrush is russet-backed and migrates to Central America for the winter, while the other populations are olive-backed and winter in South America.

russet back

more rufous in upperparts

smaller, less distinct, sparser spotting

olive-brown rump and tail

C. c. ustulatus
WESTERN

buffy eye-ring

olive-brown upperparts

buff breast

distinct blackish spots

C. c. swainsoni
EASTERN

VOICE *Single-note whit or whoit; males: several upward-spiraling phrases; flute-like song given during breeding and migration.*
NESTING *Open cup of twigs, moss, dead leaves, and mud, on branches near trunks of small trees; 3–4 eggs; 1–2 broods; Apr–Jul.*
FEEDING *Insects in flight during breeding season; berries.*
HABITAT *Coniferous forests (spruce, fir) while breeding; dense understory during migration.*
LENGTH *6½–7½in (16–19cm)*
WINGSPAN *11½–12in (29–31cm)*

Hermit Thrush

S

Catharus guttatus

This bird's song is the signature sound of northerly and mountain forests in the West—fluted, almost bi-tonal, far-carrying, and ending up with almost a question mark. The Hermit Thrush has a solitary lifestyle, especially in winter, when birds maintain inter-individual territories. Geographical variation within the vast range of the species has led to the recognition of nine subspecies (two are illustrated here).

gray-brown upperparts

C. g. faxoni
EASTERN

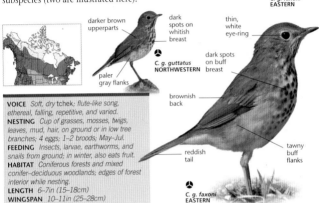

darker brown upperparts

dark spots on whitish breast

C. g. guttatus
NORTHWESTERN

paler gray flanks

thin, white eye-ring

dark spots on buff breast

brownish back

reddish tail

tawny buff flanks

C. g. faxoni
EASTERN

VOICE Soft, dry tchek; flute-like song, ethereal, falling, repetitive, and varied.
NESTING Cup of grasses, mosses, twigs, leaves, mud, hair, on ground or in low tree branches; 4 eggs; 1–2 broods; May–Jul.
FEEDING Insects, larvae, earthworms, and snails from ground; in winter, also eats fruit.
HABITAT Coniferous forests and mixed conifer–deciduous woodlands; edges of forest interior while nesting.
LENGTH 6–7in (15–18cm)
WINGSPAN 10–11in (25–28cm)

Wood Thrush

T

Hylocichla mustelina

The Wood Thrush has distinctive black spots that cover its underparts, and a rufous head and back. In the breeding season, its flute-like song echoes through the Northeastern hardwood forests and suburban forested areas. Wood Thrush populations have fallen over the past 30 years, due to habitat destruction and its susceptibility to parasitism by the Brown-headed Cowbird.

roundish, brown wings

rusty orange head and back

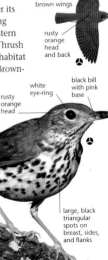

black bill with pink base

white eye-ring

rusty orange head

reddish brown lower back and rump

short, reddish brown tail

creamy pink legs and toes

large, black triangular spots on breast, sides, and flanks

VOICE Rapid pip-pippipip or rhuu-rhuu; a three-part flute-like song, ending with a trill.
NESTING Cup-shaped nest made with dried grass and weeds in trees or shrubs; 3–4 eggs; 1–2 broods; May–Jul.
FEEDING Worms, beetles, moths, caterpillars from leaf litter; fruits after breeding season.
HABITAT Interior and edges of deciduous and mixed forests while breeding; dense understory, shrubbery, and moist soil.
LENGTH 7½–8½in (19–21cm)
WINGSPAN 12–13½in (30–34cm)

American Robin

Turdus migratorius

The American Robin is the largest and most abundant of the North American thrushes, is probably the most familiar bird on the continent, and its presence on suburban lawns is an early sign of spring. It has adapted and prospered in human-altered habitats, breeding across Canada and the US and migrating out of most of Canada in fall. Migration is largely governed by changes in the availability of food, and thus, it is possible to observe some robins living off berries in suburban areas in mid-winter.

more complete
white eye-ring

♀

gray
back

orangish
red breast

dark
head

♂

white
rump

broken white
eye-ring

yellow
bill

dark
streaks
on chin

♂

mottled
gray back

dark
gray
back

spotted
breast

brick-red
underparts

fairly long,
dark tail

VOICE *High pitch tjip and multi-note, throaty tjuj-tjuk; melodious cheer-up, cheer-up, cheer-wee; sings from early morning to late evening.*
NESTING *Substantial cup of grass, weeds, and trash in tree forks or on branches, shelves, and porch lights; 4 eggs; 2–3 broods; Apr–Jul.*
FEEDING *Earthworms and small insects from leaf litter; fruit in winter.*
HABITAT *Forest, woodland, suburban gardens, parks, and farms while breeding; woodlands with berry-bearing trees.*
LENGTH 8–11in (20–28cm)
WINGSPAN 12–16in (30–41cm)

Northern Wheatear Ⓢ

Oenanthe oenanthe

The Northern Wheatear is present in North America only during its brief breeding season in Alaska and northeastern Canada. The two subspecies that breed in North America, the larger *O. o. leucorhoa* in the Northeast and *O. o. oenanthe* in the Northwest, migrate to wintering grounds in sub-Saharan Africa. The Northern Wheatear is distinguished by its black-and-white tail, which bobs when the bird walks.

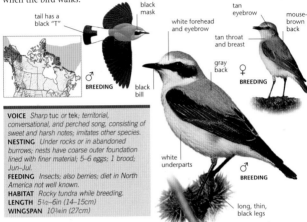

tail has a black "T"

black mask

white forehead and eyebrow

tan eyebrow

mouse-brown back

tan throat and breast

gray back ♀

BREEDING

black bill

♂ **BREEDING**

gray back

white underparts

♂ **BREEDING**

long, thin, black legs

VOICE Sharp *tuc* or *tek*; territorial, conversational, and perched song, consisting of sweet and harsh notes; imitates other species.
NESTING Under rocks or in abandoned burrows; nests have coarse outer foundation lined with finer material; 5–6 eggs; 1 brood; Jun–Jul.
FEEDING Insects; also berries; diet in North America not well known.
HABITAT Rocky tundra while breeding.
LENGTH 5½–6in (14–15cm)
WINGSPAN 10¾in (27cm)

American Dipper Ⓢ

Cinclus mexicanus

The American Dipper is at home in the cold, rushing streams of western North America. It feeds by plunging into streams for insect larvae under stones or in the streambed. When it is not foraging, it watches from a rock or log, bobbing up and down, constantly flashing its nictating membrane (the transparent third eyelid that protects the eye when the bird is underwater).

frosty scalloping on wings

pinkish bill

broad, rounded wings

paler than adult

short wings

white eyelid

straight, black bill

short tail

dark gray overall

sturdy, pink legs

VOICE Harsh *bzzt*, given singly or in rapid series; loud, disorganized series of pleasing warbles, whistles, and trills.
NESTING Domed nest with side entrance, below bridge, in bank, or behind waterfall; also provided boxes; 4–5 eggs; 1–2 broods; Mar–Aug.
FEEDING Insects and insect larvae, especially caddisflies; sometimes small fish and fish eggs.
HABITAT Fast-moving, clean mountain and coastal streams; larger, unfrozen rivers in winter.
LENGTH 7½in (19cm)
WINGSPAN 11in (28cm)

House Sparrow

S

Passer domesticus

This familiar "sparrow" is not actually a sparrow as understood in North America, but rather a member of a Eurasian family called the weaver-finches. First introduced in Brooklyn, New York, in 1850, it eventually spread across North America to the detriment of native cavity-nester songbirds. The House Sparrow has evolved to show geographic variation, being pale in the arid southwest US and darker in wetter regions.

pale rump

white wing bar ♂ ☼

gray crown

brown nape

black throat ♂ ☼

black-and-brown streaks on upperparts

gray breast

white wing bar

yellowish bill ♀

buff eyestripe

drab brown underparts

VOICE Calls include a cheery chirp, a dull jurv, and a rough jigga; chirp notes repeated endlessly.
NESTING Untidy mass of dried vegetable material in either natural or artificial cavities; 3–5 eggs; 2–3 broods; Apr–Aug.
FEEDING Mostly seeds; sometimes gleans insects and fruits.
HABITAT Downtown sections of cities and near human habitations, including agricultural areas.
LENGTH 6in (15.5cm)
WINGSPAN 9½in (24cm)

Yellow Wagtail

S

Motacilla flava

The Yellow Wagtail is a widely distributed Eurasian breeder with a nesting foothold in Alaska and the Yukon and about 17 subspecies; the Alaskan-Canadian population belongs to the subspecies *tschutschensis*. It likes to perch on exposed low shrubs and mossy mounds in the tundra, persistently wagging its tail and calling its insistent *tzeep*.

long black tail with white outer tail feathers ♂

light gray-blue head

light grayish back ♀

pale yellow underparts

dark greenish gray back

white undertail feathers

ashy blue-gray head

dark cheeks

bold, white eyebrow

bright yellow throat

inconspicuous wing bar

bright yellow belly

yellow undertail feathers

long, thin, dark legs ♂

VOICE Short, "outgoing" tzeep!; thin, musical, tzee-ouee-sir.
NESTING Small cup of woven plant matter, lined with hair and feathers; often positioned near clump of grass; 4–5 eggs; 1 brood; Jun–Jul.
FEEDING Land- and water-based insects, especially mosquitoes, along the water's edge; may catch insects in flight.
HABITAT Tundra with scattered shrub, especially along watercourses.
LENGTH 5–7in (13–18cm)
WINGSPAN 7–9in (18–23cm)

American Pipit

Anthus rubescens

In nonbreeding plumage, the American Pipit is a drab-looking bird that forages for insects along water and shores, or in cultivated fields. In the breeding season, molting transforms it into a beauty, with gray upperparts and reddish underparts. American Pipits are known for pumping their tails up and down. Breeding males display by rising into the air, then flying down with wings open and singing.

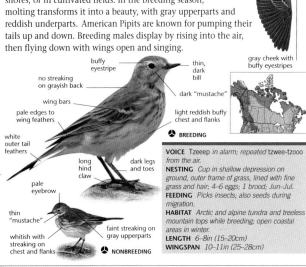

white outer tail feathers

gray cheek with buffy eyestripes

buffy eyestripe

no streaking on grayish back

thin, dark bill

dark "mustache"

wing bars

light reddish buffy chest and flanks

pale edges to wing feathers

BREEDING

white outer tail feathers

long hind claw

dark legs and toes

pale eyebrow

thin "mustache"

whitish with streaking on chest and flanks

faint streaking on gray upperparts

NONBREEDING

VOICE Tzeeep *in alarm; repeated* tzwee-tzooo *from the air.*
NESTING *Cup in shallow depression on ground, outer frame of grass, lined with fine grass and hair; 4–6 eggs; 1 brood; Jun–Jul.*
FEEDING *Picks insects; also seeds during migration.*
HABITAT *Arctic and alpine tundra and treeless mountain tops while breeding; open coastal areas in winter.*
LENGTH *6–8in (15–20cm)*
WINGSPAN *10–11in (25–28cm)*

Sprague's Pipit

Anthus spragueii

Sprague's is the only wholly North American pipit. Males perform a very extraordinary fluttering display flight, circling high while singing an unending series of high-pitched calls, for up to an hour. The current decline in the population of the Sprague's Pipit is quite likely the result of the conversion of tall-grass native prairie to extensive farmland.

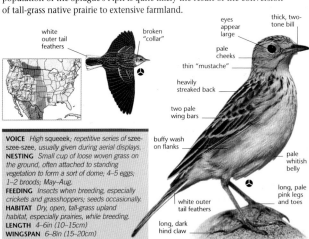

white outer tail feathers

broken "collar"

eyes appear large

thick, two-tone bill

pale cheeks

thin "mustache"

heavily streaked back

two pale wing bars

buffy wash on flanks

pale whitish belly

long, pale pink legs and toes

white outer tail feathers

long, dark hind claw

VOICE *High* squeeek; *repetitive series of* szee-szee-szee, *usually given during aerial displays.*
NESTING *Small cup of loose woven grass on the ground, often attached to standing vegetation to form a sort of dome; 4–5 eggs; 1–2 broods; May–Aug.*
FEEDING *Insects when breeding, especially crickets and grasshoppers; seeds occasionally.*
HABITAT *Dry, open, tall-grass upland habitat, especially prairies, while breeding.*
LENGTH *4–6in (10–15cm)*
WINGSPAN *6–8in (15–20cm)*

Finches

As a family of seed-eating songbirds, finches vary in size and shape from the small, fragile-looking redpolls to the robust, chunky Evening Grosbeak. Finch colors range from whitish with some pink (redpolls) to gold (American Goldfinch), bright red (crossbills), and yellow, white, and black (Evening grosbeak). They all have conical bills with razor-sharp edges, used to cut open hard seed hulls. The bills of conifer-loving crossbills are crossed at the tip, a unique arrangement that permits them to pry open tough-hulled pine cones. Crossbills wander widely to find abundant cone crops to allow breeding. Most finches are social and form flocks after nesting. All finches are vocal, calling constantly while flying, and singing in the spring. Their open cup-shaped nests of grasses and lichens hidden in trees or shrubs are hard to find.

GARDEN GLOW
Even pink flower buds cannot compete with the brilliant yellow of a male American Goldfinch.

American Goldfinch

S

Spinus tristis

Often described as a giant yellow-and-black bumblebee, a male American Goldfinch in full breeding plumage is a common summer sight. They are one of summer's latest breeding birds, awaiting the arrival of milkweed and thistle seeds to build their nests. Even when not seen, the presence of goldfinches in an area is quickly given away by the sound of the birds calling in flight. They relish eating weed seeds and will readily come to nyjer seed feeders in the backyard. Males perform courtship songs, often while circling the females.

brownish bill

tan back

yellow throat and collar

bright yellow back

pale tan underparts

♂ **NONBREEDING**

♂ **NONBREEDING**

black forehead and crown

black tail

short conical pinkish bill

white rump

white wing bar

bright yellow underparts

♂ **BREEDING**

pinkish legs and feet

brownish olive back

pinkish bill

brownish overall

dull yellow throat

♀ **BREEDING**

♀ **NONBREEDING**

VOICE *Males: loud, rising pter-yee; tit-tse-tew-tew by both sexes, usually in flight; complex warbling.*
NESTING *Open cup nest of grass and/or thistleseed, usually shaded from above; 4–5 eggs; 1–2 broods; Jul–Sep.*
FEEDING *Seeds from annual plants, birch, and alder; some insects; prefers sunflower and thistle seed at feeders.*
HABITAT *Low shrubs, deciduous woodlands, farmlands, orchards, suburbs, and gardens.*
LENGTH *4¼–5in (11–13cm)*
WINGSPAN *7–9in (18–23cm)*

Pine Siskin

S

Spinus pinus

This energetic, fearless little bird of the conifer belt runs in gangs, zipping over the trees with incessant twittering. An expert at disguise, the Pine Siskin can resemble a cluster of pine needles or cones. Often abundant wherever there are pines, spruces, and other conifers, they may still make a mass exodus from a region if the food supply is not to their liking.

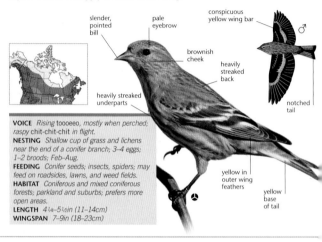

slender, pointed bill

pale eyebrow

conspicuous yellow wing bar ♂

brownish cheek

heavily streaked back

heavily streaked underparts

notched tail

yellow in outer wing feathers

yellow base of tail

VOICE *Rising toooeeo, mostly when perched; raspy chit-chit-chit in flight.*
NESTING *Shallow cup of grass and lichens near the end of a conifer branch; 3–4 eggs; 1–2 broods; Feb–Aug.*
FEEDING *Conifer seeds; insects, spiders; may feed on roadsides, lawns, and weed fields.*
HABITAT *Coniferous and mixed coniferous forests; parkland and suburbs; prefers more open areas.*
LENGTH *4¼–5½in (11–14cm)*
WINGSPAN *7–9in (18–23cm)*

Common Redpoll

S

Acanthis flammea

Every other year, spruce, birch, and other trees in the northern forest zone fail to produce a good crop of seeds, forcing the Common Redpoll to look for food farther south than usual. It is tame around people and easily attracted to winter feeders. The degree of whiteness in its plumage varies greatly among individuals, due to sex and age.

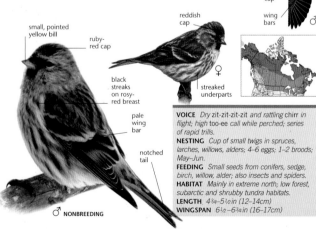

red cap

wing bars ♂

small, pointed yellow bill

ruby-red cap

reddish cap

♀

black streaks on rosy-red breast

streaked underparts

pale wing bar

notched tail

VOICE *Dry zit-zit-zit-zit and rattling chirr in flight; high too-ee call while perched; series of rapid trills.*
NESTING *Cup of small twigs in spruces, larches, willows, alders; 4–6 eggs; 1–2 broods; May–Jun.*
FEEDING *Small seeds from conifers, sedge, birch, willow, alder; also insects and spiders.*
HABITAT *Mainly in extreme north; low forest, subarctic and shrubby tundra habitats.*
LENGTH *4¾–5½in (12–14cm)*
WINGSPAN *6½–6¾in (16–17cm)*

♂ NONBREEDING

Hoary Redpoll

Acanthis hornemanni

This bird of the high Arctic has two recognized subspecies—*A. h. exilipes* and *A. h. hornemanni*. Hoary Redpolls are close relatives of the Common Redpoll and often breed in the same areas. Like Common Redpolls, their chattering flocks buzz rapidly over trees and fields, and while they are tame around humans, this species is less well known because of its more limited contact with people.

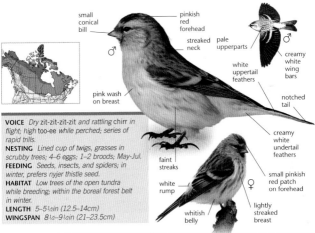

small conical bill

pinkish red forehead

streaked neck

pale upperparts

♂

creamy white wing bars

white uppertail feathers

notched tail

pink wash on breast

♂

creamy white undertail feathers

faint streaks

white rump

small pinkish red patch on forehead

♀

lightly streaked breast

whitish belly

VOICE Dry *zit-zit-zit-zit* and rattling *chirr* in flight; high *too-ee* while perched; series of rapid trills.
NESTING Lined cup of twigs, grasses in scrubby trees; 4–6 eggs; 1–2 broods; May–Jul.
FEEDING Seeds, insects, and spiders; in winter, prefers nyjer thistle seed.
HABITAT Low trees of the open tundra while breeding; within the boreal forest belt in winter.
LENGTH 5–5½in (12.5–14cm)
WINGSPAN 8½–9½in (21–23.5cm)

Gray-crowned Rosy-Finch

Leucosticte tephrocotis

High western mountains are the domain of these brown-and-pink birds, which are seldom seen. The Gray-crowned Rosy-Finch is one of several mountain finch species that extend across the Bering Strait into eastern Asia. Geographically variable in size and coloration, with exceptionally large forms occurring on the Pribilof and Aleutian Islands, the Gray-crowned Rosy-Finch is the most abundant of the North American rosy-finches.

brown upperparts

🐦 WESTERN

rosy rump

dark crown

gray behind eye

more extensive gray on head

dark chin

🐦 WESTERN

pinkish shoulder patch

notched tail

rusty brown underparts

whitish undertail feathers

🐦 INTERIOR

VOICE High-pitched *peew*, given singly or in short series; repetitive series of *twee* notes.
NESTING Bulky assemblage of grasses, lichens, and twigs in cracks or under rock overhangs; 3–5 eggs; 1–2 broods; May–Jun.
FEEDING Seeds from the ground; insects and their larvae in summer.
HABITAT Alpine habitats such as rocky screes above snow line and tundra while breeding; lower elevations in winter.
LENGTH 5½–8½in (14–21cm)
WINGSPAN 13in (33cm)

Purple Finch

Carpodacus purpureus

The Purple Finch is more often seen than heard, even on their breeding grounds in open and mixed coniferous forest. The western subspecies (*californicus*) is slightly darker and duller than the eastern form (*purpuleus*). Not to be confused with the strawberry-colored House Finches, the raspberry-red male Purple Finches are easier to identify than the brown-streaked females.

pinkish red body

♂

round, brownish wings

raspberry red crown

♂

brown stripe between eye and bill

pink and brown streaked upperparts

whitish belly with rosy patches

pink rump and upper tail

lightly streaked overall

darker, streaked wings

brownish, conical bill

pale brown overall

♀

VOICE *Single, rough pikh in flight; rich series of notes, up and down in pitch.*
NESTING *Cup of sticks and grasses on a conifer branch; 4 eggs; 2 broods; May–Jul.*
FEEDING *Buds, seeds, flowers of deciduous trees; insects and caterpillars in summer; also seeds and berries; feeding stations, especially in winter.*
HABITAT *Northern mixed conifer and hardwood forests while breeding.*
LENGTH *4¾–6in (12–15cm)*
WINGSPAN *8½–10in (22–26cm)*

Cassin's Finch

Carpodacus cassinii

This finch has a melodious song that incorporates phrases from several different Rocky Mountain species. From below, the male Cassin's Finch resembles a sparrow, but when it alights, its purple-red plumage is evident. This species closely resembles the Purple and House Finches, whose ranges it overlaps, so accurate identification may be difficult. The female Cassin's Finch resembles a generic fledgling or a sparrow.

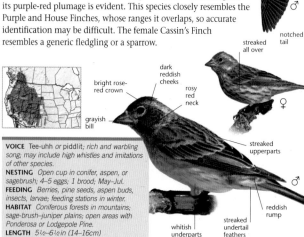

red face

♂

notched tail

streaked all over

dark reddish cheeks

bright rose-red crown

rosy red neck

grayish bill

♀

streaked upperparts

♂

reddish rump

whitish underparts

streaked undertail feathers

VOICE *Tee-uhh or piddlit; rich and warbling song; may include high whistles and imitations of other species.*
NESTING *Open cup in conifer, aspen, or sagebrush; 4–5 eggs; 1 brood; May–Jul.*
FEEDING *Berries, pine seeds, aspen buds, insects, larvae; feeding stations in winter.*
HABITAT *Coniferous forests in mountains; sage-brush–juniper plains; open areas with Ponderosa or Lodgepole Pine.*
LENGTH *5½–6½in (14–16cm)*
WINGSPAN *10–10½in (25–27cm)*

House Finch

Carpodacus mexicanus

Historically, the House Finch was a western bird, and was first reported in the East on Long Island, New York City in 1941. These birds are said to have originated from released pets from the illegal bird trade. The population of the eastern birds started expanding in the 1960s; by the late 1990s, the eastern and western populations had linked up. The male House Finch is distinguished from the Purple and Cassin's finches by its brown streaked underparts and more strawberry-colored wash, while the females have plainer faces and generally blurrier streaking.

pinkish head

red face

brown cap

♂ NONBREEDING

♂ BREEDING

grayish streaks all over

usually brick-red bib and head

brown upperparts

♀

pale brown streaking

streaked belly

long tail feathers

brown streaked undertail feathers

♂ BREEDING

VOICE Queet; *jumble of notes, often starting with husky notes to whistled and burry notes, ending with a long* wheeerr.
NESTING *Nest of grass stems, thin twigs, and thin weeds in trees and on man-made structures; 1–6 eggs; 2–3 broods; Mar–Aug.*
FEEDING *Vegetable matter, such as buds, fruits, and seeds; readily comes to feeders.*
HABITAT *Urban, suburban, and settled areas; wilder areas in West, such as savannas and desert grasslands.*
LENGTH 5–6in (12.5–15cm)
WINGSPAN 8–10in (20–25cm)

Red Crossbill

Loxia curvirostra

The Crossbill's highly adapted bill is used to bite between the scales of a conifer cone and pry them apart, then the seed is lifted out with its tongue. Red Crossbills are further specialized to harvest conifer seeds; eight different forms have been recognized, with different body sizes, bill shapes, and sizes. Each form has a different flight call and rarely interbreeds with other forms.

crown usually brick-red

♂

black stripe over eye

dark wings

crossed mandibles

greenish breast

♀

dark brown wings

red body

♂

black wings

VOICE Jit repeated 2–5 times; complex, continuous warbling of notes, whistles, and buzzes.
NESTING Cup nest on lateral conifer branch; 3–5 eggs; 2 broods; can breed year-round.
FEEDING Pine seeds; also insects and larvae, particularly aphids; other seeds.
HABITAT Coniferous or mixed-coniferous and deciduous forests; mountain forests in the Rockies.
LENGTH 5–6¾in (13–17cm)
WINGSPAN 10–10½in (25–27cm)

red rump

White-winged Crossbill

Loxia leucoptera

Few other creatures of the northerly forest go about their business with such determined energy as the White-winged Crossbill, and no others accent a winter woodland with hot pink and magenta. Flocks of these birds gather in spruce trees, calling with a chorus of metallic, yanking notes, then erupt into the air.

variable dark patch on cheek

♂

two conspicuous white wing bars

brownish green head

dark brown wings

greenish streaked underparts

♀

crossed mandibles

blackish wings

red body

♂

VOICE Sharp, chattering plik, or deeper tyoop, repeated in series of 3–7 notes; song melodious trilling.
NESTING Open cup nest, usually high on end of a spruce branch; 3–5 eggs; 2 broods; Jul, Jan–Feb.
FEEDING Seeds from small-coned conifers; spruces, firs, larches; insects when available.
HABITAT Nomadic; coniferous forests, especially with spruce and tamarack.
LENGTH 5½–6in (14–15cm)
WINGSPAN 10–10½in (26–27cm)

pinkish red underparts

notched tail

Pine Grosbeak ⓢ

Pinicola enucleator

The largest member of the finch family in North America, and easily distinguished by the male's unmistakable thick, stubby bill, and pink color, the Pine Grosbeak favors boreal forests from coast to coast and readily attends feeders. Four subspecies are found in North America. Due to extensive color variation of individual plumages, the age and sex of the bird are not always easily determined.

stubby, curved, blackish bill
pinkish-red head
short neck
pinkish red underparts (but regionally variable)
two white wing bars
pinkish rump
long, blackish tail
greenish head
greenish rump
pale patch under eye
gray belly

VOICE Contact calls of eastern birds tee-tew, or tee-tee-tew; western forms give more complex tweedle; warbling song.
NESTING Well-hidden, open cup nest in spruce or larch trees; 2–5 eggs, 1 brood; Jun–Jul.
FEEDING Spruce buds, maple seeds, and mountain ash berries throughout the year; feeding stations in winter; insects in summer.
HABITAT Open, northerly coniferous forests usually near fresh water.
LENGTH 8–10in (20–25cm)
WINGSPAN 13in (33cm)

Evening Grosbeak ⓢ

Coccothraustes vespertinus

The husky Evening Grosbeak feeds in noisy flocks at bird feeders in the winter. It has extended its range eastward in the past 200 years and now nests as far as Newfoundland, perhaps due to the planting of ornamental box elder, whose seeds ensure a ready food supply in winter. Recently though, its eastern range is rescinding northward, possible due to spruce budworm control measures.

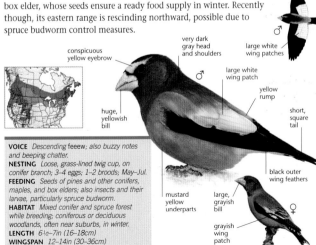

black wing tips
very dark gray head and shoulders
conspicuous yellow eyebrow
large white wing patches
large white wing patch
yellow rump
short, square tail
huge, yellowish bill
black outer wing feathers
mustard yellow underparts
large, grayish bill
grayish wing patch

VOICE Descending feeew; also buzzy notes and beeping chatter.
NESTING Loose, grass-lined twig cup, on conifer branch; 3–4 eggs; 1–2 broods; May–Jul.
FEEDING Seeds of pines and other conifers, maples, and box elders; also insects and their larvae, particularly spruce budworm.
HABITAT Mixed conifer and spruce forest while breeding; coniferous or deciduous woodlands, often near suburbs, in winter.
LENGTH 6½–7in (16–18cm)
WINGSPAN 12–14in (30–36cm)

Wood-warblers

Wood-warblers are remarkable for their diversity in plumage, song, feeding, breeding biology, and sexual dimorphism. In general, though, warblers share similar shapes: all are smallish birds with longish, thin bills used mostly for snapping up invertebrates. The odd, chunky, thick-billed Yellow-breasted Chat is a notable exception, but genetic data suggest what many birders have long suspected: it's not a warbler at all. Ground-dwelling warblers tend to be larger and clad in olives, browns, and yellows, while many arboreal species are small and sport bright oranges, cool blues, and even ruby reds. The color, location, and presence or absence of paler wingbars and tail spots is often a good identification aid.

STATIC PLUMAGE
Some warblers, such as this male Golden-winged Warbler, keep their stunning plumage year-round.

Golden-winged Warbler

Vermivora chrysoptera

This species is unfortunately being genetically swamped by the Blue-winged Warbler. It commonly interbreeds with the Blue-winged, resulting in two more frequently seen hybrid forms: Brewster's Warbler, which resembles the Blue-winged Warbler, and Lawrence's Warbler, which looks like a Blue-winged Warbler with the mask and black throat of a Golden-winged.

white outer tail feathers

bright yellow wing panel

gray back

gray back suffused with yellow

unstreaked wings

black "mask"

bright yellow crown

black throat

yellow wing panel

white undertail

gray "mask"

greenish yellow crown

VOICE Sharp tsip; high, slightly buzzy ziiih in flight; buzzy zee zuu zuu zuu, first note higher.
NESTING Shallow bulky cup, on or just above ground; 4–6 eggs; 1 brood; May–Jul.
FEEDING Moth larvae, other winged insects, and spiders; hangs upside-down at clusters of curled-up dead leaves.
HABITAT Short secondary growth habitat with dense deciduous shrubs, or marshes with forest edge while breeding.
LENGTH 4¾in (12cm)
WINGSPAN 7½in (19cm)

Blue-winged Warbler

Vermivora pinus

The Blue-winged Warbler breeds along forest edges and in second growth. It interbreeds freely with Golden-winged Warblers, producing a variety of fertile combinations. The most frequently produced hybrid, Brewster's Warbler, is similar to the Golden-winged Warbler (yellowish breast, two yellow wing bars), but has the Blue-winged's facial pattern, minus the black mask and throat.

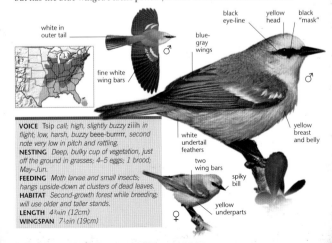

white in outer tail

black eye-line

yellow head

black "mask"

blue-gray wings

fine white wing bars

white undertail feathers

two wing bars

spiky bill

yellow underparts

yellow breast and belly

VOICE Tsip call; high, slightly buzzy ziiih in flight; low, harsh, buzzy beee-burrrrr, second note very low in pitch and rattling.
NESTING Deep, bulky cup of vegetation, just off the ground in grasses; 4–5 eggs; 1 brood; May–Jun.
FEEDING Moth larvae and small insects; hangs upside-down at clusters of dead leaves.
HABITAT Second-growth forest while breeding; will use older and taller stands.
LENGTH 4¾in (12cm)
WINGSPAN 7½in (19cm)

Tennessee Warbler ⓢ

Vermivora peregrina

The Tennessee Warbler was named after its place of
discovery, but this bird would have been on migration, as
it breeds almost entirely in Canada and winters in Central
America. These warblers inhabit fairly remote areas, and
their nests are difficult to find. The population of
Tennessee Warblers tends to increase in years when
budworms are abundant.

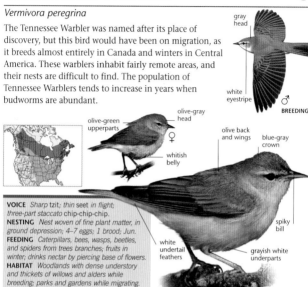

gray
head

white
eyestripe

♂ **BREEDING**

olive-green
upperparts

olive-gray
head

♀

whitish
belly

olive back
and wings

blue-gray
crown

spiky
bill

white
undertail
feathers

grayish white
underparts

VOICE *Sharp tzit; thin seet in flight;*
three-part staccato chip-chip-chip.
NESTING *Nest woven of fine plant matter, in*
ground depression; 4–7 eggs; 1 brood; Jun.
FEEDING *Caterpillars, bees, wasps, beetles,*
and spiders from trees branches; fruits in
winter; drinks nectar by piercing base of flowers.
HABITAT *Woodlands with dense understory*
and thickets of willows and alders while
breeding; parks and gardens while migrating.
LENGTH 4¾in (12cm)
WINGSPAN 7¾in (19.5cm)

♂ **BREEDING**

Orange-crowned Warbler ⓢ

Vermivora celata

Common and relatively brightly colored in the West but uncommon and duller in
the East, the Orange-crowned Warbler has a large breeding range. The
19th-century American naturalist Thomas Say described this species based on
specimens collected in Nebraska. Because its tiny orange cap is concealed in the
plumage of the crown, he named it *celata*, which is Latin for "hidden."

dull olive
overall

♂

greenish
yellow
rump

olive-green
upperparts

short
wings

crown shows
orange when
bird alarmed

pale
yellow
eyebrow

▲ **WEST**

muted
breast
markings

gray
head

drabber
plumage
overall

yellow
undertail
feathers

◐ **EAST; 1ST** ❄

VOICE *Clean, sharp tsik; high, short seet*
in flight; loose, lazy trill.
NESTING *Cup of grasses, fibers, and down,*
usually on ground under bush; 4–5 eggs;
1 brood; Mar–Jul.
FEEDING *Arthropods such as beetles, ants,*
spiders, and their larvae; also fruits; drinks
nectar by piercing base of flower.
HABITAT *Varied habitats while breeding;*
prefers streamside thickets.
LENGTH 5in (13cm)
WINGSPAN 7¼in (18.5cm)

Nashville Warbler

Vermivora ruficapilla

Although often confused with the ground-walking Connecticut Warbler, the Nashville Warbler is smaller, hops about in trees, and has a yellow throat. Nashville has two subspecies: *V. r. ruficapilla* in the East and *V. r. ridgwayi* in the West. Differences in voice, habitat, behavior, and plumage hint that they may be separate species. *V. r. ridgwayi* has more white on its belly and a grayish green back.

conspicuous white eye-ring
rufous crown patch
blue-gray helmet
grayish green back
♂ *V. r. ruficapilla* EASTERN
olive-green upperparts
olive wings
rounded wings

♂ *V. r. ridgwayi* WESTERN

yellow undertail feathers
white patch on belly
duller olive back

less contrast between gray and yellow
♀ *V. r. ruficapilla* EASTERN

VOICE Sharp tik; high, thin siit in flight; eastern song: first lazy, then faster trill tee-tsee tee-tsee tee-tsee titititi; western song: slightly lower with a seldom trilled tee-tsee tee-tsee tee-tsee weesay weesay way.
NESTING Cup hidden on ground in dense cover; 3–6 eggs; 1 brood; May–Jul.
FEEDING Insects and spiders from trees.
HABITAT Wet habitats in East and brushy montane areas in West while breeding.
LENGTH 4¾in (12cm)
WINGSPAN 7½in (19cm)

Northern Parula

Parula americana

The Northern Parula is a small wood-warbler that somewhat resembles a chickadee in its active foraging behavior. This bird depends on very specific nesting materials—*Usnea* lichens, or "Old Man's Beard," in the north, and *Tillandsia*, or Spanish Moss, in the South—greatly limiting the geographical range of this species.

two white wing bars
♂

dark patch between eye and bill
yellow throat
interrupted white eye-ring
blue-gray neck and head
chestnut streaks on chest
olive back
♂

♀
yellow chest, lacks chestnut streaks

gray rump and uppertail
delicate, pale gray belly
dark legs

VOICE Very sharp tsip; thin, weak, descending tsiif in flight; buzzy upslurred trill, ending high, then dropping off in an emphatic zip.
NESTING Hanging pouch in clump of lichens; 4–5 eggs; 1 brood; Apr–Aug (north).
FEEDING Caterpillars, flies, moths, beetles, wasps, ants, spiders; also berries, nectar, some seeds.
HABITAT Wooded areas with preferred nesting material while breeding.
LENGTH 4¼in (11cm)
WINGSPAN 7in (18cm)

white patches on outer tail feathers
pinkish yellow feet

Yellow Warbler

S

Dendroica petechia

By May, the song of the Yellow Warbler can be heard across North America as the birds arrive for the summer. This species is extremely variable geographically, with about 40 subspecies, especially on its tropical range. It is known to build another nest on top of an old one when cowbird eggs appear in it, which can result in up to six different tiers. The Yellow Warbler does not walk, but rather hops from branch to branch.

dark flight feathers with yellow edges

♂

faint yellow wing bars

bright yellow face with conspicuous black eye

thin, pointed bill

yellow upperparts

♂

rusty streaks on breast and flanks

mostly yellow tail

dull brown legs and toes

dull yellowish overall

plain face

yellowish olive back

♀

♂ ♀
1ST ✳

yellow underparts

VOICE Variable chip, sometimes in series; buzzy zeep flight call; variable series of fast, sweet notes; western birds often add an emphatic ending.
NESTING Deep cup of plant material, grasses in vertical fork of deciduous tree or shrub; 4–5 eggs; 1 brood; May–Jul.
FEEDING Mostly insects and insect larvae from the tops of tall shrubs and small trees; some fruit.
HABITAT Widespread in shrubby and second-growth habitats.
LENGTH 5in (13cm)
WINGSPAN 8in (20cm)

Chestnut-sided Warbler Ⓢ

Dendroica pensylvanica

The Chestnut-sided Warbler depends on deciduous second-growth and forest edges for breeding. These birds vary in appearance, immature females looking quite unlike adult males in breeding. In all plumages, yellowish wing bars and whitish belly are the most distinguishing characteristics. Its pleasant song has long been transcribed as *pleased pleased pleased to MEET'cha.*

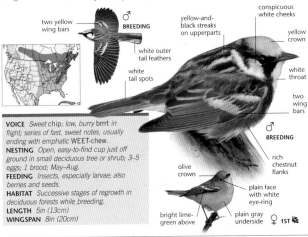

two yellow wing bars

♂ BREEDING

white outer tail feathers

white tail spots

conspicuous white cheeks

yellow-and-black streaks on upperparts

yellow crown

white throat

two wing bars

♂ BREEDING

rich chestnut flanks

olive crown

plain face with white eye-ring

bright lime-green above

plain gray underside ♀ 1ST

VOICE *Sweet chip; low, burry brrrt in flight; series of fast, sweet notes, usually ending with emphatic WEET-chew.*
NESTING *Open, easy-to-find cup just off ground in small deciduous tree or shrub; 3–5 eggs; 1 brood; May–Aug.*
FEEDING *Insects, especially larvae; also berries and seeds.*
HABITAT *Successive stages of regrowth in deciduous forests while breeding.*
LENGTH *5in (13cm)*
WINGSPAN *8in (20cm)*

Blackpoll Warbler Ⓢ

Dendroica striata

The Blackpoll Warbler is known for undergoing a remarkable fall migration that takes it over the Atlantic Ocean from southern Canada and the northeastern US to northern Venezuela. Before departing, it almost doubles its body weight with fat to serve as fuel for the nonstop journey. In spring, most of these birds travel the shorter Caribbean route back north.

white tail spots

♂

two white wing bars

greenish upperparts with fine black streaks

♀ BREEDING

faint, fine streaking on underparts

black cap

white cheek

bold black streaks on gray back

streaked underparts

white undertail feathers

orange legs

♂ BREEDING

VOICE *Piercing chip; high, buzzy, sharp tzzzt in flight; crescendo of fast, extremely high-pitched ticks, ending with a decrescendo tsst tsst TSST TSST TSST tsst tsst.*
NESTING *Well-hidden cup placed low against conifer trunk; 3–5 eggs; 1–2 broods; May–Jul.*
FEEDING *Arthropods, such as worms and beetles; small fruit in fall and winter.*
HABITAT *Spruce-fir forests and coastal coniferous forests while breeding.*
LENGTH *5½in (14cm)*
WINGSPAN *9in (23cm)*

Bay-breasted Warbler (S)

Dendroica castanea

Male Bay-breasted Warblers in breeding plumage are striking birds, but fall females are very different with their dull, greenish plumage. It depends largely on outbreaks of spruce budworms (a major food source), so its numbers rise and fall according to those outbreaks. Overall, the Bay-breasted Warbler population has decreased because of the increased use of pesticide sprays.

chestnut crown, streaked black

two white wing bars

white tips on outer tail feathers

bold buffy neck patch

buffy wash on flanks and under tail

dusky ear patch

♀ BREEDING

♂ BREEDING

gray upperparts with black streaks

chestnut brown crown

chestnut brown chin and flanks

two white wing bars

buff undertail

yellowish buff belly

black face

♂ BREEDING

VOICE *Upslurred* tsip; *high, buzzy* tzzzt; *high, thin* wee-si wee-si wee-si wee, *ending on lower pitch.*
NESTING *Fragile-looking cup of grass and lichens on horizontal branch at mid-level in forest; 4–5 eggs; 1 brood; May–Jul.*
FEEDING *Moths, smaller insects, worms, spiders, caterpillars while breeding; fruit in winter.*
HABITAT *Mature spruce-fir forest while breeding; varied habitats during migration.*
LENGTH *5½in (14cm)*
WINGSPAN *9in (23cm)*

Blackburnian Warbler (S)

Dendroica fusca

The Blackburnian Warbler's orange throat is unique among the North American warblers. It co-exists with many other *Dendroica* warblers in the coniferous and mixed woods of the north and east, but is able to do so by exploiting a slightly different niche for foraging—in this case the treetops. It also seeks the highest trees for nesting.

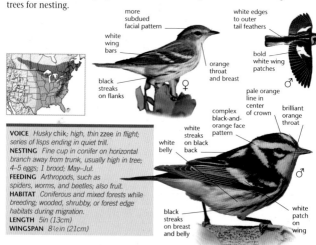

more subdued facial pattern

white edges to outer tail feathers

white wing bars

orange throat and breast

bold white wing patches

black streaks on flanks

♀

♂

pale orange line in center of crown

complex black-and-orange face pattern

brilliant orange throat

white streaks on black back

white belly

♂

black streaks on breast and belly

white patch on wing

VOICE *Husky* chik; *high, thin* zzee *in flight; series of lisps ending in quiet trill.*
NESTING *Fine cup in conifer on horizontal branch away from trunk, usually high in tree; 4–5 eggs; 1 brood; May–Jul.*
FEEDING *Arthropods, such as spiders, worms, and beetles; also fruit.*
HABITAT *Coniferous and mixed forests while breeding; wooded, shrubby, or forest edge habitats during migration.*
LENGTH *5in (13cm)*
WINGSPAN *8½in (21cm)*

Magnolia Warbler

Ⓢ

Dendroica magnolia

The bold, flashy, and common Magnolia Warbler is hard to miss as it flits around at eye level, fanning its uniquely marked tail. This species nests in young forests and winters in almost any habitat, so its numbers have not suffered in recent decades, unlike some of its relatives. Although it really has no preference for its namesake plant, the 19th-century ornithologist Alexander Wilson discovered one feeding in a magnolia tree during migration, which is how it got its name.

yellow rump

gray crown

♂ **BREEDING**

broken white tail band

plain face with pale eye-ring

greenish back

white undertail feathers

greenish back with black stripes

black streaking on breast and flanks not as heavy

white eyebrow

incomplete eye-ring

♀ **BREEDING**

black face

large white patch on wing

♂ **BREEDING**

yellow underparts with black streaks

VOICE Tinny *jeinf, not particularly warbler-like; short, simple whistled series* wee'-sa wee'-sa WEET-a-chew; *high, trilled* zeep *in flight, short and distinctive.*
NESTING *Flimsy cup of black rootlets placed low in dense conifer against trunk; 3–5 eggs; 1 brood; Jun–Aug.*
FEEDING *Mostly caterpillars, beetles, and spiders from the underside of conifer needles and broadleaf foilage.*
HABITAT *Dense, young mixed and coniferous forests while breeding.*
LENGTH *5in (13cm)*
WINGSPAN *7½in (19cm)*

Cape May Warbler ⓢ

Dendroica tigrina

The Cape May Warbler is a spruce budworm specialist; its population increases during outbreaks of that insect. These birds often chase other birds aggressively from flowering trees, where they use their thin, pointed bills and semi-tubular tongues to suck the nectar from blossoms. It also uses its bill to feed on insects by plucking them from clumps of conifer needles.

white patches on wings ♂

pale yellow nape

black cap

gray back

yellow nape

thin, pointed bill

rufous cheeks

yellow underparts, heavily streaked with black

white patches on flanks and breast

♀

white marks on outer tail feathers

♂

VOICE *High, even-pitched series of whistles* see see see see.
NESTING Cup placed near trunk, high in spruce or fir; 4–9 eggs; 1 brood; Jun–Jul.
FEEDING Arthropods, especially spruce budworms; flies, moths, and beetles from mid-high levels in canopy; fruit and nectar during the nonbreeding season.
HABITAT Mature spruce–fir forests while breeding; varied habitats during migration.
LENGTH 5in (13cm)
WINGSPAN 8in (20cm)

Black-throated Blue Warbler ⓢ

Dendroica caerulescens

Male and female Black-throated Blue Warblers look so dissimilar that they were originally considered different species. Many of the females have a blue wash to their wings and tail, and almost all have a subdued version of the male's white "kerchief." These warblers migrate northward in spring along the Appalachians, but a small number of birds fly northwestward to the Great Lakes.

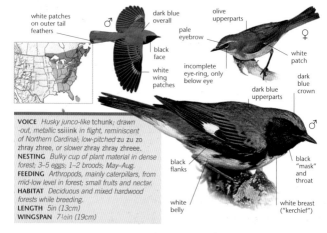

white patches on outer tail feathers ♂

dark blue overall

olive upperparts

pale eyebrow

♀

white patch

black face

white wing patches

incomplete eye-ring, only below eye

dark blue upperparts

dark blue crown

♂

black "mask" and throat

black flanks

white belly

white breast ("kerchief")

VOICE *Husky junco-like tchunk; drawn-out, metallic ssiiink in flight, reminiscent of Northern Cardinal; low-pitched zu zu zo zhray zhree, or slower zhray zhray zhreee.*
NESTING Bulky cup of plant material in dense forest; 3–5 eggs; 1–2 broods; May–Aug.
FEEDING Arthropods, mainly caterpillars, from mid-low level in forest; small fruits and nectar.
HABITAT Deciduous and mixed hardwood forests while breeding.
LENGTH 5in (13cm)
WINGSPAN 7½in (19cm)

Yellow-rumped Warbler

Ⓢ

Dendroica coronata

The abundant and widespread Yellow-rumped Warbler is not choosy
about its wintering habitats. It was once considered to consist of
two species, "Myrtle" (*D. c. coronata*) in the East, and "Audubon's"
(*D. c. auduboni*) in the West. Because they interbreed freely in a
narrow zone of contact in British Columbia and Alberta, the
American Ornithologists Union merged them. The two forms
differ in plumage and voice, and their hybrid zone appears stable.

white wing bars

♂ MYRTLE

♀ *D. c. auduboni* AUDUBON'S

yellowish throat

grayish overall

dark cheeks

black streaks on gray back

white throat

♂ *D. c. coronata* MYRTLE

black streaks across breast

bright yellow rump

white corners on outer tail feathers

lacks white eyebrow

whitish eyebrow

whitish throat

same pattern as male, but duller

yellow flanks

♀ *D. c. coronata* MYRTLE

solid black breast

large, white wing patch

♂ *D. c. auduboni* AUDUBON'S

unmarked undertail

VOICE coronata: *flat, husky tchik;* auduboni: *higher-pitched, rising*
jip; both: clear, upslurred sviiit *in flight; loose, warbled trill with an*
inflected ending.
NESTING *Bulky cup of plant matter in conifer; 4–5 eggs; 1 brood;*
Mar–Aug.
FEEDING *Flies, beetles, wasps, and spiders while breeding; fruit*
and berries at other times of the year; often sallies to catch prey.
HABITAT *Coniferous and mixed hardwood coniferous forests.*
LENGTH *5in (13cm)*
WINGSPAN *9in (23cm)*

Black-throated Gray Warbler ⓢ

Dendroica nigrescens

The Black-throated Gray Warbler inhabits the understory of forests and woodlands of oak and mixed woodlands in dry to arid western North America. It has a leisurely foraging style, and its nest is built by both males and females, placed only a feet few from the ground. These birds linger in their range until late fall, sometimes even wintering in California and Arizona.

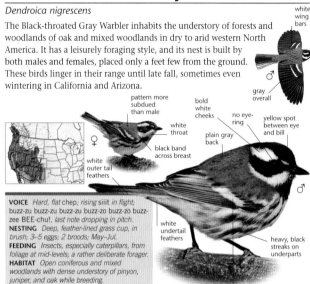

white wing bars

♂

gray overall

pattern more subdued than male

bold white cheeks

white throat

no eye-ring

yellow spot between eye and bill

plain gray back

♀

black band across breast

white outer tail feathers

♂

white undertail feathers

heavy, black streaks on underparts

one or two white wing bars

VOICE Hard, flat chep; rising siiit in flight; buzz-zu buzz-zu buzz-zu buzz-zo buzz-zo buzz-zee BEE-chu!, last note dropping in pitch.
NESTING Deep, feather-lined grass cup, in brush; 3–5 eggs; 2 broods; May–Jul.
FEEDING Insects, especially caterpillars, from foliage at mid-levels; a rather deliberate forager.
HABITAT Open coniferous and mixed woodlands with dense understory of pinyon, juniper, and oak while breeding.
LENGTH 5in (13cm)
WINGSPAN 7½in (19cm)

Townsend's Warbler ⓢ

Dendroica townsendi

The buzzy song of Townsend's Warbler can be heard from the treetops in the Northwest's fir forests between May and July. Birds from the Pacific Northwest winter along the Pacific coast. The Townsend's Warbler interbreeds with the closely related Hermit Warbler in Washington and Oregon, resulting in varied-looking hybrids.

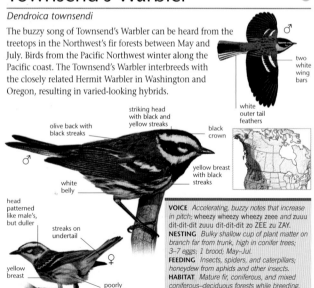

♂

two white wing bars

white outer tail feathers

striking head with black and yellow streaks

black crown

olive back with black streaks

♂

yellow breast with black streaks

white belly

head patterned like male's, but duller

streaks on undertail

♀

yellow breast

poorly marked streaks on flanks

VOICE Accelerating, buzzy notes that increase in pitch; wheezy wheezy wheezy zeee and zuuu dit-dit-dit zuuu dit-dit-dit zo ZEE zu ZAY.
NESTING Bulky shallow cup of plant matter on branch far from trunk, high in conifer trees; 3–7 eggs; 1 brood; May–Jul.
FEEDING Insects, spiders, and caterpillars; honeydew from aphids and other insects.
HABITAT Mature fir, coniferous, and mixed coniferous–deciduous forests while breeding.
LENGTH 5in (13cm)
WINGSPAN 8in (20cm)

Black-throated Green Warbler (S)

Dendroica virens

This species is easy to distinguish as its bright yellow face is unique among birds inhabiting northeastern North America. It is a member of the *virens* "superspecies," a group of non-overlapping species that are similar in plumage and vocalizations—the Black-throated Green, Golden-cheeked, Townsend's, and Hermit Warblers. Sadly, this species is vulnerable to habitat loss in parts of its wintering range.

greenish cap

two white wing bars

olive-green back

yellow face

♂

white outer tail feathers

black bib and chin

heavily streaked underparts

yellowish flanks

same as male, but duller

greenish flanks

♀

VOICE *Flat tchip; rising siii in flight; fast zee zee zee zee zoo zee; slower zu zee zu-zu zee.*
NESTING *Cup of twigs and grasses high on branch near trunk; 3–5 eggs; 1 brood; May–Jul.*
FEEDING *Arthropods, especially caterpillars; also small fruit, including poison ivy berries, in nonbreeding season.*
HABITAT *Forests, especially mix of conifers and hardwood while breeding; variety of habitats during migration and winter.*
LENGTH *5in (13cm)*
WINGSPAN *8in (20cm)*

Pine Warbler (S)

Dendroica pinus

Pine Warblers live in pine forests in southeastern Canada and the eastern US; it is often the most common bird in its namesake habitat and its distinctive song can be heard from several birds at once. One of the few warblers that visits feeders, the Pine Warbler is a hardy bird, staying within the US throughout the winter.

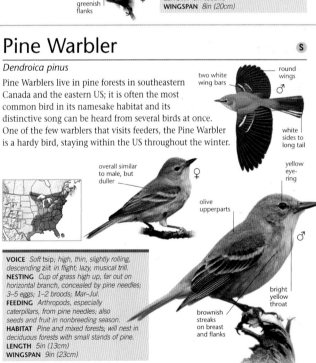

two white wing bars

round wings

♂

white sides to long tail

overall similar to male, but duller

♀

yellow eye-ring

olive upperparts

♂

bright yellow throat

brownish streaks on breast and flanks

VOICE *Soft tsip; high, thin, slightly rolling, descending ziit in flight; lazy, musical trill.*
NESTING *Cup of grass high up, far out on horizontal branch, concealed by pine needles; 3–5 eggs; 1–2 broods; Mar–Jul.*
FEEDING *Arthropods, especially caterpillars, from pine needles; also seeds and fruit in nonbreeding season.*
HABITAT *Pine and mixed forests; will nest in deciduous forests with small stands of pine.*
LENGTH *5in (13cm)*
WINGSPAN *9in (23cm)*

Palm Warbler

(S)

Dendroica palmarum

The Palm Warbler is one of North America's most abundant warblers. Its tail-pumping habit makes it easy to identify in any plumage. The western subspecies (*D. p. palmarum*), found in Western and Central Canada, is grayish brown above and lacks the chestnut streaks of the eastern subspecies (*D. p. hypochrysea*), which has a yellower face and breeds in southeastern Canada and northeastern US.

chestnut crown

white-edged tail

dark upperparts

EASTERN

yellow eyestripe

grayish green "mustache"

dull gray upperparts

yellow undertail feathers

ring below eye

dark gray upperparts

yellow throat

chestnut streaks on breast

rich yellow underparts

D. p. hypochrysea EASTERN; BREEDING

yellowish rump

dusky streaks on breast and belly

D. p. palmarum WESTERN; BREEDING

VOICE Husky chik or tsip; light ziint in flight; slow, loose, buzzy trill: zwi zwi zwi zwi zwi zwi zwi zwi.
NESTING Cup of grasses on or near ground in open area of conifers at forest edge of a bog; 4–5 eggs; 1 brood; May–Jul.
FEEDING Insects, sometimes caught in flight; seeds and berries.
HABITAT Spruce bogs within northerly forest zone while breeding.
LENGTH 5½in (14cm)
WINGSPAN 8in (20cm)

Black-and-white Warbler

(S)

Mniotilta varia

The Black-and-white Warbler is best known for its creeper-like habit of feeding in vertical and upside-down positions as it pries into bark crevices, where its relatively long, curved bill allows it to reach tiny nooks and crannies. It is a long-distance migrant, with some birds wintering in parts of northern South America.

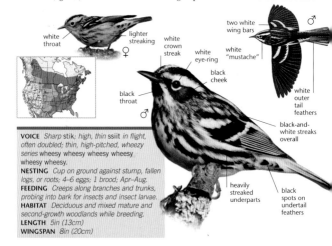

white throat

lighter streaking ♀

white crown streak

two white wing bars ♂

white eye-ring

white "mustache"

black cheek

black throat ♂

white outer tail feathers

black-and-white streaks overall

heavily streaked underparts

black spots on undertail feathers

VOICE Sharp stik; high, thin ssiit in flight, often doubled; thin, high-pitched, wheezy series wheesy wheesy wheesy wheesy wheesy wheesy.
NESTING Cup on ground against stump, fallen logs, or roots; 4–6 eggs; 1 brood; Apr–Aug.
FEEDING Creeps along branches and trunks, probing into bark for insects and insect larvae.
HABITAT Deciduous and mixed mature and second-growth woodlands while breeding.
LENGTH 5in (13cm)
WINGSPAN 8in (20cm)

Cerulean Warbler

Setophaga cerulean

This unusually colored species is difficult to spot, as it spends the majority of its time foraging in the canopy of deciduous forests. It was once common across the Midwest and the Ohio River Valley, but its habitat is being cleared for agriculture and development. In winter, this bird lives high in the canopy of the Andean foothills, a habitat threatened by coffee cultivation.

two white wing bars

short tail with white band

whitish eyebrow

pale blue crown

sea-green upperparts

yellowish underparts

♀

blue upperparts

white undertail feathers

bright blue crown

indistinct eyestripe

♂

white chin and throat

black breastband

black streaks on flanks

white belly

VOICE *Slurred chip; buzzy zeet in flight; short series of low, buzzy, paired notes followed by a mid-range trill and upslurred high-pitched zhree.*
NESTING *Compact cup high on fork in deciduous tree, far from trunk; 2–5 eggs; 1 brood; May–Jul.*
FEEDING *Insects from high in canopy.*
HABITAT *Mature deciduous forests while breeding; dense woods during migration.*
LENGTH *4¾in (12cm)*
WINGSPAN *7¾in (19.5cm)*

Prairie Warbler

Setophaga discolor

Despite its name, the Prairie Warbler does not live on the "prairie." Its distinctive song is a quintessential sound of scrubby areas in its range. Although the population of this bird increased in the 19th century due to the widespread clearing of forests, the maturation of this habitat, along with human development, is having a negative impact on some populations.

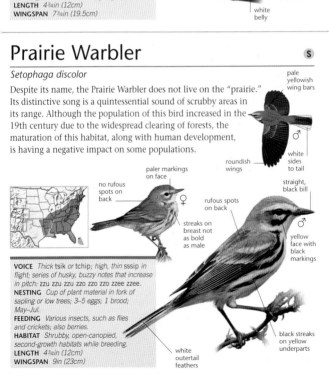

pale yellowish wing bars

♂

white sides to tail

roundish wings

paler markings on face

no rufous spots on back

♀

rufous spots on back

streaks on breast not as bold as male

straight, black bill

♂

yellow face with black markings

black streaks on yellow underparts

white outertail feathers

VOICE *Thick tsik or tchip; high, thin sssip in flight; series of husky, buzzy notes that increase in pitch: zzu zzu zzu zzo zzo zzo zzee zzee.*
NESTING *Cup of plant material in fork of sapling or low trees; 3–5 eggs; 1 brood; May–Jul.*
FEEDING *Various insects, such as flies and crickets; also berries.*
HABITAT *Shrubby, open-canopied, second-growth habitats while breeding.*
LENGTH *4¾in (12cm)*
WINGSPAN *9in (23cm)*

American Redstart

Setophaga ruticilla

The American Redstart is a vividly colored, energetic and acrobatic warbler with a reasonably broad range across North America. One of its behavioral quirks is to fan its tail and wings while foraging, supposedly using the flashes of bold color to scare insects into moving, making them easy prey. It possesses well-developed rictal bristles, hair-like feathers extending from the corners of the mouth, which help it to "feel" insects.

yellow tail base

olive back

grayish head

♀

conspicuous orange wing bar

yellowish flanks

whitish underparts

black inverted "T" on tail ♂

yellow tail base

♂ ☻

irregular, dark patches

yellow flanks

black head and back

♂

long, black tail with orange on sides

blackish smudge on undertail

orange flank patch with black border

white belly

VOICE Harsh tsiip; high, thin sweep in flight; song variable, high, thin, penetrating series of notes; burry, emphatic, and downslurred see-a see-a see-a ZEE-urrrr.
NESTING Cup of grasses and rootlets, lined with feathers; placed low in deciduous tree; 2–5 eggs; 1–2 broods; May–Jul.
FEEDING Insects and spiders from leaves at mid-levels in trees; also moths, flies in flight; fruit.
HABITAT Moist deciduous and mixed woodlands while breeding.
LENGTH 5in (13cm)
WINGSPAN 8in (20cm)

Hooded Warbler

S

Setophaga citrina

The Hooded Warbler is a strikingly patterned and loud warbler. Both male and females frequently flash the white markings hidden on the inner webs of their tails. The extent of the black hood varies in female Hooded Warblers; it ranges from none in first fall birds to almost as extensive as males in some adult females.

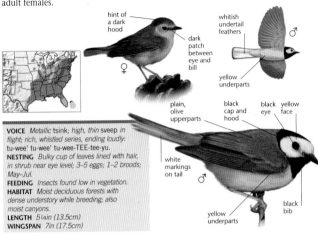

hint of a dark hood

♀

dark patch between eye and bill

whitish undertail feathers

♂

yellow underparts

plain, olive upperparts

black cap and hood

black eye

yellow face

white markings on tail

♂

black bib

yellow underparts

VOICE Metallic tsink; high, thin sweep in flight; rich, whistled series, ending loudly: tu-wee' tu-wee' tu-wee-TEE-tee-yu.
NESTING Bulky cup of leaves lined with hair, in shrub near eye level; 3–5 eggs; 1–2 broods; May–Jul.
FEEDING Insects found low in vegetation.
HABITAT Moist deciduous forests with dense understory while breeding; also moist canyons.
LENGTH 5¼in (13.5cm)
WINGSPAN 7in (17.5cm)

Prothonotary Warbler

Protonotaria citrea

The ringing song of the Prothonotary Warbler echoes through the swampy forests of southern Ontario and southeastern US every summer. This is one of the few cavity-nesting warbler species; it will use manmade bird houses placed close to still water. Prothonotary Warblers also tend to stay fairly low over the water, making them easy to spot.

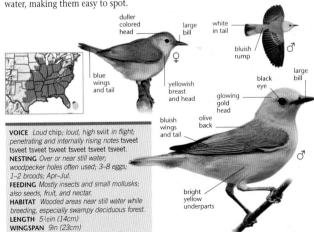

duller colored head

large bill

white in tail

♂

bluish rump

♀

blue wings and tail

yellowish breast and head

black eye

large bill

glowing gold head

bluish wings and tail

olive back

♂

bright yellow underparts

VOICE Loud chip; loud, high sviit in flight; penetrating and internally rising notes tsveet tsveet tsveet tsveet tsveet tsveet tsveet.
NESTING Over or near still water; woodpecker holes often used; 3–8 eggs; 1–2 broods; Apr–Jul.
FEEDING Mostly insects and small mollusks; also seeds, fruit, and nectar.
HABITAT Wooded areas near still water while breeding, especially swampy deciduous forest.
LENGTH 5½in (14cm)
WINGSPAN 9in (23cm)

Ovenbird

(S)

Seiurus aurocapilla

The Ovenbird is so-called for the domed, oven-like nests it builds on the ground in mixed and deciduous forests. Males flit about boisterously, often at night, incorporating portions of their main song into a jumble of spluttering notes. In the forest, one male singing loudly to declare his territory can set off a whole chain of responses from his neighbors.

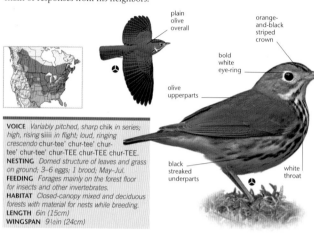

plain olive overall

orange-and-black striped crown

bold white eye-ring

olive upperparts

black streaked underparts

white throat

VOICE Variably pitched, sharp chik in series; high, rising siiii in flight; loud, ringing crescendo chur-tee' chur-tee' chur-tee' chur-tee' chur-TEE chur-TEE chur-TEE.
NESTING Domed structure of leaves and grass on ground; 3–6 eggs; 1 brood; May–Jul.
FEEDING Forages mainly on the forest floor for insects and other invertebrates.
HABITAT Closed-canopy mixed and deciduous forests with material for nests while breeding.
LENGTH 6in (15cm)
WINGSPAN 9½in (24cm)

Northern Waterthrush

(S)

Seiurus noveboracensis

The tail-bobbing Northern Waterthrush is often heard giving a spink! call as it swiftly flees from observers. Although this species may be mistaken for the closely related Louisiana Waterthrush, the Northern Waterthrush can be distinguished by its song and its preference for still over running water.

pale eyebrow

streaking on white or yellowish flanks

dull brown upperparts

pale eyebrow narrows behind eye

small, short bill

short tail

dull, fleshy-colored legs and toes

fine, dense breast streaking

VOICE Sharp, rising, ringing spink!; rising, buzzy ziiiit in flight; loud, accelerating, staccato notes teet, teet, toh-toh toh-toh tyew-tyew!
NESTING Hair-lined, mossy cup, in tree roots or riverbank; 4–5 eggs; 1 brood; May–Aug.
FEEDING Insects such as ants, mosquitoes, moths, and beetles; slugs and snails; small crustaceans and tiny fish when migrating.
HABITAT Still-water swamps and bogs while breeding; also still edges of rivers and lakes.
LENGTH 6in (15cm)
WINGSPAN 9½in (24cm)

Louisiana Waterthrush

D

Parkesia motacilla

The Louisiana Waterthrush is one of the earliest warblers to return north in the spring; as early as March, eastern ravines are filled with cascades of its song. Both the stream-loving Louisiana Waterthrush and its still-water cousin, the Northern Waterthrush, bob their tails as they walk, but the Louisiana Waterthrush arcs its entire body at the same time.

short tail

dull brown overall

buffy area near bill and eye

unstreaked throat

bicolored flanks; white forward, washed cinnamon on rear

white eyebrow flares behind eye

large bill

thick, sparse breast streaking

bright, bubble gum pink legs and toes in spring

VOICE *Round spink; rising ziiiit in flight; whistles, ending with* see'-oh see'-oh see'-uh see'-uh tip-uh-tik-uh-tik-uh-tip-whee'ur-tik.
NESTING *Bulky mass of leaves, moss, and twigs; under steep stream bank over water; 4–6 eggs; 1 brood; May–Aug.*
FEEDING *Insect larvae, snails, small fish from streams; flying insects (dragonflies, stoneflies).*
HABITAT *Fast-moving streams in deciduous forests while breeding.*
LENGTH *6in (15cm)*
WINGSPAN *10in (25cm)*

Connecticut Warbler

S

Oporornis agilis

The shy Connecticut Warbler breeds in remote, boggy habitats in Canada and is hard to spot during its spring and fall migrations. It arrives in the US in late May and leaves its breeding grounds in August. It is the only warbler that walks along the ground in a bouncy manner, with its tail bobbing up and down.

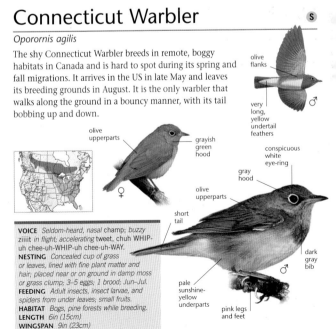

olive flanks

very long, yellow undertail feathers

♂

olive upperparts

grayish green hood

♀

conspicuous white eye-ring

gray hood

olive upperparts

short tail

dark gray bib

♂

pale sunshine-yellow underparts

pink legs and feet

VOICE *Seldom-heard, nasal champ; buzzy ziiiit in flight; accelerating tweet, chuh WHIP-uh chee-uh-WHIP-uh chee-uh-WAY.*
NESTING *Concealed cup of grass or leaves, lined with fine plant matter and hair; placed near or on ground in damp moss or grass clump; 3–5 eggs; 1 brood; Jun–Jul.*
FEEDING *Adult insects, insect larvae, and spiders from under leaves; small fruits.*
HABITAT *Bogs, pine forests while breeding.*
LENGTH *6in (15cm)*
WINGSPAN *9in (23cm)*

Mourning Warbler ⑤

Oporornis philadelphia

The Mourning Warbler's song is often used in movies as a
background sound of idyllic suburban settings. However, this
warbler is unlikely to be found in a backyard, as it prefers
dense, herbaceous tangles for breeding and during migration.
These birds are late spring migrants and the leaves are fully out
when they arrive in the East, making them difficult to see.

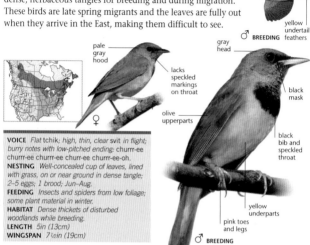

"hooded" look

yellow undertail feathers

♂ **BREEDING**

gray head

pale gray hood

lacks speckled markings on throat

olive upperparts

♀

black mask

black bib and speckled throat

yellow underparts

pink toes and legs

♂ **BREEDING**

VOICE *Flat tchik; high, thin, clear svit in flight;
burry notes with low-pitched ending: churrr-ee
churrr-ee churrr-ee churr-ee churrr-ee-oh.*
NESTING *Well-concealed cup of leaves, lined
with grass, on or near ground in dense tangle;
2–5 eggs; 1 brood; Jun–Aug.*
FEEDING *Insects and spiders from low foliage;
some plant material in winter.*
HABITAT *Dense thickets of disturbed
woodlands while breeding.*
LENGTH *5in (13cm)*
WINGSPAN *7½in (19cm)*

MacGillivray's Warbler ⑤

Oporornis tolmiei

This western counterpart to the Mourning Warbler is
distinguished from it by the incomplete white eye-ring. Even
though John Kirk Townsend had originally named this bird
"Tolmie's Warbler" after Dr. W.T. Tolmie, John James Audubon
later called it MacGillivray's warbler honoring his editor and
the name stuck.

incomplete, white eye-ring

♂

yellow undertail feathers

grayish head

olive upperparts

♀

pale gray throat

olive upperparts

incomplete eye-ring

gray head

black patch between eye and bill

♂

irregular, black bib

yellow underparts

pinkish legs and toes

VOICE *Sharp tssik; high, thin, clear svit in
flight; loud, staccato, rolling series; ends lower
or higher than rest of song.*
NESTING *Cup of plant material just off the
ground in deciduous shrubs and thickets; 3–5
eggs; 1 brood; May–Aug.*
FEEDING *Beetles, flies, bees, caterpillars from
low foliage, just above ground level.*
HABITAT *Thickets in mixed and coniferous
forests while breeding, often along streams.*
LENGTH *5in (13cm)*
WINGSPAN *7½in (19cm)*

Common Yellowthroat

Geothlypis trichas

This common warbler is noticeable partly because of its loud, simple song. This species varies in voice and plumage across its range and 14 subspecies have been described. Western populations have yellower underparts, brighter white head stripes, and louder, simpler songs than the eastern birds. The male often flies upwards rapidly, delivering a more complex version of its song.

black mask

♂

plain, olive-green overall

black "mask" including forehead

pale stripe over "mask," varies from gray to white or yellowish

pale eye-ring

olive upperparts

yellow throat

yellow throat

♀

olive-green upperparts

♂

greenish gray underparts

olive-green tail

VOICE *Harsh, buzzy tchak, repeated into chatter when agitated; low, flat, buzzy dzzzit in flight; rich (often three-note) phrases: WITCH-uh-tee WITCH-uh-tee WITCH-uh-tee WHICH.*
NESTING *Concealed, bulky cup of grasses just above ground; 3–5 eggs; 1 brood; May–Aug.*
FEEDING *Insects and spiders from low vegetation; also seeds.*
HABITAT *Dense herbaceous understory, in marshes, grasslands, forests, hedgerows.*
LENGTH *5in (13cm)*
WINGSPAN *6¾in (17cm)*

Wilson's Warbler

Wilsonia pusilla

The tiny Wilson's Warbler is perhaps the most common spring migrant of all the wood-warblers across many areas of the West. In the East, however, it is much scarcer in spring. Wilson's Warblers have a wide range of habitats, yet their numbers are declining, especially in the West, as its riverside breeding habitats are gradually being destroyed by development.

♂

long, narrow tail

olive or blackish crown

♀

yellow eyebrow and chin

black cap

large black eye

olive upperparts

♂

yellow brightest on face

VOICE *Rich chimp or champ; sharp, liquid tsik in flight; chattering trill, often increasing in speed che che che chi-chi-chi-chit.*
NESTING *Cup of leaves and grass placed on or near ground in mosses or grass; 4–6 eggs; 1 brood; Apr–Jun.*
FEEDING *Insects from foliage, leaf litter, or during flight; berries and honeydew.*
HABITAT *Wet shrubby thickets with no canopy while breeding, often near streams and lakes.*
LENGTH *4¾in (12cm)*
WINGSPAN *7in (17.5cm)*

Canada Warbler

Cardellina canadensis

One of the last species of wood-warblers to arrive in Canada in the spring, and among the first to leave in the fall, the Canada Warbler is recognizable by the conspicuous black markings on its chest. This bird is declining, probably because of the maturation and draining of its preferred breeding habitat of old mixed hardwood forests with moist undergrowth.

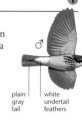

plain gray tail

white undertail feathers

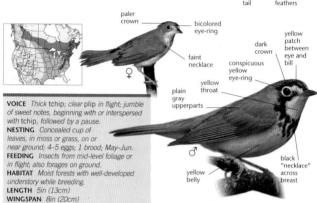

paler crown

bicolored eye-ring

faint necklace

yellow patch between eye and bill

dark crown

conspicuous yellow eye-ring

plain gray upperparts

yellow throat

black "necklace" across breast

yellow belly

VOICE Thick *tchip*; clear *plip* in flight; jumble of sweet notes, beginning with or interspersed with *tchip*, followed by a pause.
NESTING Concealed cup of leaves, in moss or grass, on or near ground; 4–5 eggs; 1 brood; May–Jun.
FEEDING Insects from mid-level foliage or in flight; also forages on ground.
HABITAT Moist forests with well-developed understory while breeding.
LENGTH 5in (13cm)
WINGSPAN 8in (20cm)

Yellow-breasted Chat

Icteria virens

This species, the largest of all warblers in Canada, has puzzled ornithologists for a long time, as DNA studies have given conflicting results about whether it actually belongs to the wood-warbler family or not. Sometimes, it skulks in dense vegetation; at other times, it sits singing atop small trees. One of its behavioral quirks is to suddenly fly upwards, then glide slowly back to earth, while singing.

rounded wings

yellow underwing feathers

duller olive upperparts

buff patch between eye and bill

black patch between eye and bill

thick, blackish bill

white "spectacles"

olive upperparts

long, rounded tail

bright yellow breast

black legs and feet

VOICE Seldom-heard, low, soft *tuk* and nasal, downslurred *tiyew*; monosyllabic grunts, clucks, and whistles; sometimes sings at night; mimics other birds.
NESTING Concealed and bulky structure of dead plant matter, in thicket near eye-level; 3–5 eggs; 1–2 broods; May–Aug.
FEEDING Insects; fruit and berries.
HABITAT Dense shrubby areas, forest edges while breeding; thickets along rivers in West.
LENGTH 7½in (19cm)
WINGSPAN 9½in (24cm)

Orioles & Blackbirds

Members of this diverse family of birds are common and widespread, occurring from coast to coast in nearly every habitat in North America.

Brightly colored orioles build intricate hanging nests that are an impressive combination of engineering and weaving. Most species boast a melodious song and tolerance for humans.

Cowbirds are parasitic birds that lay eggs in the nests of close to 300 different species in the Americas. The species found in Canada has a thick bill and blackish, iridescent body contrasting with a brown head.

Blackbirds are largely covered in dark feathers, and their long, pointed bills and tails add to their streamlined appearance. They are among the most numerous birds on the continent after the breeding season.

Meadowlarks are birds of the open country. They can be recognized by a characteristic bright-yellow chest with a black V-shaped bib and a sweet singing voice.

NECTAR LOVER
A Baltimore Oriole inserts its bill into the base of a flower, taking the nectar.

Orchard Oriole

Ⓢ

Icterus spurius

The Orchard Oriole resembles a large warbler in size, color, and the way it flits among leaves while foraging for insects. Unlike other orioles, it flutters its tail, and spends less time on the breeding grounds, often arriving there as late as mid-May and leaving as early as late July. It tolerates humans and can be found breeding in suburban parks and gardens.

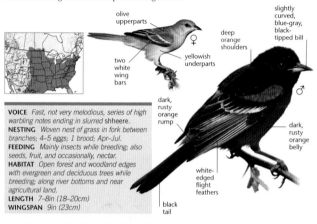

black back

olive upperparts

♀

two white wing bars

yellowish underparts

slightly curved, blue-gray, black-tipped bill

deep orange shoulders

dark, rusty orange rump

♂

dark, rusty orange belly

white-edged flight feathers

black tail

VOICE Fast, not very melodious, series of high warbling notes ending in slurred *shheere.*
NESTING Woven nest of grass in fork between branches; 4–5 eggs; 1 brood; Apr–Jul.
FEEDING Mainly insects while breeding; also seeds, fruit, and occasionally, nectar.
HABITAT Open forest and woodland edges with evergreen and deciduous trees while breeding; along river bottoms and near agricultural land.
LENGTH 7–8in (18–20cm)
WINGSPAN 9in (23cm)

Bullock's Oriole

Ⓢ

Icterus bullockii

The Bullock's Oriole is the western counterpart of the Baltimore in both behavior and habitat. The two were thought to belong to a single species called the Northern Oriole (*L. galbula*), because they interbreed where they overlap in the Great Plains. Recent studies, however, suggest that they are separate species. The Bullock's Oriole is more resistant to brood parasites and punctures and removes cowbird eggs from its nest.

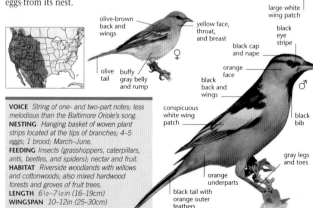

♂

large white wing patch

olive-brown back and wings

yellow face, throat, and breast

♀

olive tail

buffy gray belly and rump

black cap and nape

black eye stripe

orange face

black back and wings

♂

black bib

conspicuous white wing patch

gray legs and toes

orange underparts

black tail with orange outer feathers

VOICE String of one- and two-part notes; less melodious than the Baltimore Oriole's song.
NESTING Hanging basket of woven plant strips located at the tips of branches; 4–5 eggs; 1 brood; March–June.
FEEDING Insects (grasshoppers, caterpillars, ants, beetles, and spiders); nectar and fruit.
HABITAT Riverside willows and cottonwoods; also mixed hardwood forests and groves of fruit trees.
LENGTH 6½–7½in (16–19cm)
WINGSPAN 10–12in (25–30cm)

Baltimore Oriole

Icterus galbula

The Baltimore Oriole's brilliant colors are familiar to many in eastern North America because this bird is so tolerant of humans. This species originally favored the American Elm for nesting, but Dutch Elm disease decimated these trees. The oriole has since adapted to using sycamores, cottonwoods, and other tall trees as nesting sites. Its ability to use suburban gardens and parks has helped expand its range to incorporate areas densely occupied by humans.

orange-yellow head

♂1ST

white-edged black wings

♂

black and orange tail

orange-yellow shoulder patch

yellow-olive rump

olive upperparts

two wing bars

pale orange underparts

♀

straight blue-gray bill

black head

black back

♂

black upper breast

orange rump

black tail with orange outer tail feathers

orange underparts

VOICE *Loud, clear, melodious song comprising several short notes in series, often of varying lengths.*
NESTING *Round-bottomed basket usually woven of grass, hung toward the end of branches; 4–5 eggs; 1 brood; May–Jul.*
FEEDING *Hops or flits among leaves and branches picking insects and spiders; caterpillars; also fruit and nectar.*
HABITAT *Forest edges and tall, open mixed hardwoods, close to rivers; forested parks, urban areas with tall trees.*
LENGTH *8–10in (20–26cm)*
WINGSPAN *10–12in (26–30cm)*

Brown-headed Cowbird ⓢ

Molothrus ater

North America's most common brood parasite, the Brown-headed Cowbird was once a bird of the Great Plains, following bison to prey on insects kicked up by their hooves. Now it is found continent-wide. It is a serious threat to North American songbirds, laying its eggs in the nests of more than 220 different species, and having its young raised to fledging by more than 140 species.

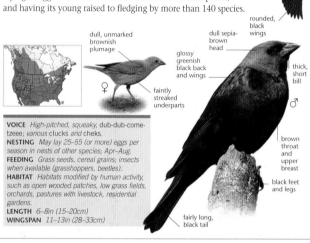

♂

rounded, black wings

dull sepia-brown head

glossy greenish black back and wings

thick, short bill

♂

dull, unmarked brownish plumage

♀

faintly streaked underparts

brown throat and upper breast

black feet and legs

fairly long, black tail

VOICE High-pitched, squeaky, dub-dub-come-tzeee; various clucks and cheks.
NESTING May lay 25–55 (or more) eggs per season in nests of other species; Apr–Aug.
FEEDING Grass seeds, cereal grains; insects when available (grasshoppers, beetles).
HABITAT Habitats modified by human activity, such as open wooded patches, low grass fields, orchards, pastures with livestock, residential gardens.
LENGTH 6–8in (15–20cm)
WINGSPAN 11–13in (28–33cm)

Red-winged Blackbird ⓢ

Agelaius phoeniceus

One of the most abundant native bird species in North America, the Red-winged Blackbird is conspicuous in wetland habitats, even in suburbia. The sight and sound of males singing from the tops of cattails is a sure sign that spring is near. This adaptable species migrates and roosts in huge flocks. There are numerous subspecies, one of the most distinctive being the "Bicolored" Blackbird (*A. p. gubernator*).

♂

red and yellow "flags"

black outer wings

black eye

pointed bill

all-black back and tail

bright red shoulder patches with yellow edge

♂

light brown eyebrow

♀

off-white underparts with dark streaks

VOICE Various brusk chek, chit, or chet; males: kronk-a-rhee with a characteristic nasal, rolling and metallic "undulating" ending.
NESTING Cup of grasses and mud in reeds or cattails; 3–4 eggs; 1–2 broods; Mar–Jun.
FEEDING Seeds and grains; largely insects when breeding; feeders.
HABITAT Wetlands, especially freshwater marshes; wet meadows with tall grass, and open woodlands with reedy vegetation.
LENGTH 7–10in (18–25cm)
WINGSPAN 11–14in (28–35cm)

Rusty Blackbird

short, narrow bill

♂
BREEDING

long tail

Euphagus carolinus

The Rusty Blackbird breeds in remote, inaccessible swampy areas, and is much less of a pest to agricultural operations than some of the other members of its family. The plumage on the male Rusty Blackbird changes to a dull, reddish brown during the fall. This species is most easily observed during fall migration, as it moves south in long, wide flocks.

green sheen on head

pale whitish or yellow eye

gray-brown eyebrow

pale gray to rusty brown underparts

♀

black overall, with blue-green to greenish sheen

♂
BREEDING

VOICE Chuk *during migration flights; males: musical* too-ta-lee.
NESTING *Small bowl of branches and sticks, lined with wet plants and dry grass, usually near water; 3–5 eggs; 1 brood; May–Jul.*
FEEDING *Seasonally available insects, spiders, grains, seeds of trees, and fleshy fruits or berries.*
HABITAT *Moist to wet forests up to the northern timberline while breeding.*
LENGTH *8–10in (20–25cm)*
WINGSPAN *12–15in (30–38cm)*

Brewer's Blackbird

stout bill

♂

long, dark tail

Euphagus cyanocephalus

The Brewer's Blackbird seems to prefer areas disturbed by humans to natural ones throughout much of its range. Interestingly, when the Brewer's Blackbird range overlaps with that of the Common Grackle, it wins out in rural areas, but loses out in urban areas. This species can be found feasting on waste grains left behind after the harvest.

brown eyes

gray brown overall

♀

purplish sheen on head

yellow eyes

black body with greenish blue sheen

♂

black legs and feet

VOICE *Buzzy* tshrrep *song ascending in tone.*
NESTING *Bulky cup of dry grass, stem and twig framework lined with soft grasses and animal hair; 3–6 eggs; 1–2 broods; Apr–Jul.*
FEEDING *Insects from the ground during breeding season; snails; seeds, grain, and occasional fruit in fall and winter.*
HABITAT *Open, disturbed areas and human development including parks, gardens, cleared forests, and fallow fields.*
LENGTH *10–12in (25–30cm)*
WINGSPAN *13–16in (33–41cm)*

Common Grackle

Quiscalus quiscula

The Common Grackle is so well suited to urban and suburban habitats that it successfully excludes other species from them and even kills and eats House Sparrows. During migration and winter, Common Grackles form immense flocks, sometimes comprising more than 1 million individuals. This tendency, combined with its preference for cultivated areas, has made this species an agricultural pest in some regions.

dark wings

pale yellow eye

iridescent bluish purple head

iridescent brownish bronze back

long, thick bill

♂ BRONZED FORM

long, v-shaped tail

dull purplish bronze overall

pale eye

♀

VOICE *Low, harsh chek; loud series of odd squeaks and whistles.*
NESTING *Small bowl in trees, with a frame of sticks filled with mud and grasses; 4–6 eggs; 1–2 broods; Apr–Jul.*
FEEDING *Beetles, flies, spiders, and worms; small vertebrates; seeds and grain, especially in nonbreeding season.*
HABITAT *Open woodlands, suburban woodlots, city parks, gardens, and hedgerows.*
LENGTH *11–13½in (28–34cm)*
WINGSPAN *15–18in (38–46cm)*

Eastern Meadowlark

Sturnella magna

A bird of the eastern grassy fields, the colorful Eastern Meadowlark is known for its plaintive sounding song. During courtship, the male sings enthusiastically from the highest available perch. This species overlaps with the very similar looking Western Meadowlark in the western Great Plains and are best distinguished by their different calls and songs.

black-and-white striped crown

rounded wings

buffy wash on face

buffy mottling in black breastband

long, pointed bill

black stripe behind eye

whitish face

brown upperparts streaked with buff and black

yellow throat

yellow breast with black "v"

yellow belly

short tail with white outer tail feathers

yellow belly

long toes

BREEDING

VOICE *Sharp dzzeer; clear, descending, 3–8 note whistles, tseeeooou tseeeeou.*
NESTING *Loosely woven, usually domed, cup of grasses and other plants, located on the ground in tall grass fields; 3–8 eggs; 1 brood; Mar–May.*
FEEDING *Insects from ground (grasshoppers, caterpillars, grubs); seeds, grain in winter.*
HABITAT *Native tallgrass openings, pastures, and overgrown roadsides while breeding.*
LENGTH *7–10in (18–25cm)*
WINGSPAN *13–15in (33–38cm)*

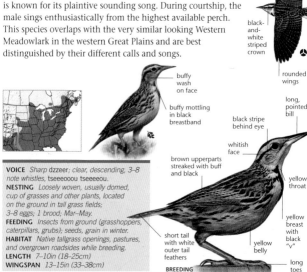

Western Meadowlark

Sturnella neglecta

Although the range of the Western Meadowlark overlaps widely with that of its eastern counterpart, hybrids between the two species are very rare and usually sterile. The large numbers of Western Meadowlarks in the western Great Plains, the Great Basin, and the Central Valley of California, combined with the male's tendency to sing conspicuously from the tops of shrubs, when fence posts are not available, make this species attractive to birdwatchers. Where the two meadowlarks overlap, they are best identified by their song.

duller pattern than breeding bird

short wings

white outer tail feathers

yellow throat

yellow patch between bill and eye

NONBREEDING

blackish brown stripe behind eye

long, pointed bill

chunky body

black "v" on yellow chest

black spots and streaks on sides and flanks

yellow underparts

BREEDING

short, wide tail

long toes

VOICE *Series of complex, bubbling, whistled notes descending in pitch.*
NESTING *Domed grass cup, well hidden in tall grasses; 3–7 eggs; 1 brood; Mar–Aug.*
FEEDING *Insects, including beetles, grubs, and grasshoppers; also grains and grass seeds.*
HABITAT *Open grassy plains while breeding; also agricultural fields with overgrown edges and hayfields.*
LENGTH *7–10in (18–26cm)*
WINGSPAN *13–15in (33–38cm)*

Yellow-headed Blackbird (S)

Xanthocephalus xanthocephalus

The male with its conspicuous bright yellow head is unmistakable; females are more drab. Populations of the Yellow-headed Blackbird fluctuate widely but locally according to available rainfall, which controls the availability and quality of its breeding marshland habitat. In some wetlands, these birds are extremely abundant and easily noticeable due to their amazing song.

yellow head

conspicuous white wing patches

brownish overall

yellowish throat and facial patch ♀

bright yellow head and chest

black overall

white wing patch ♂

long tail

VOICE *Nasal whaah; series of harsh, cackling noises, followed by a brief pause, and a high, long, wailing trill.*
NESTING *Cup of plant strips woven into standing aquatic vegetation; 3–4 eggs; 1 brood; May–Jun.*
FEEDING *Insects while breeding; agricultural grains and grass seeds in winter.*
HABITAT *Marshes with cattails and bulrushes while breeding; wetlands with wooded areas.*
LENGTH *8½–10½in (21–27cm)*
WINGSPAN *15in (38cm)*

Bobolink (T)

Dolichonyx oryzivorus

The Bobolink is a common summer resident of open fallow fields through much of the northern US and southern Canada. In spring, the males perform a conspicuous circling or "helicoptering" display while singing, to establish territory and attract females. Bobolink populations have declined on both the breeding grounds and in wintering areas because of habitat loss and haying practices during nesting.

black wings

♂ BREEDING

buff-colored hindneck

blackish brown crown

pinkish bill

gold-buff overall

♀ BREEDING

white shoulder feathers

black face and crown

black underparts

white rump

♂ BREEDING

black tail with pointed feathers

VOICE *Link; long, complex babbling series of musical notes varying in length and pitch.*
NESTING *Woven cup of grass close to or on the ground, well hidden in tall grass; 3–7 eggs; 1 brood; May–Jul.*
FEEDING *Mostly insects, spiders, grubs in breeding season; cereal grains and grass seeds.*
HABITAT *Open fields with mixture of tall grasses and other herbaceous vegetation with breeding; especially old hayfields.*
LENGTH *6–8in (15–20cm)*
WINGSPAN *10–12in (25–30cm)*

American Sparrows

Many North American birds in this diverse family have shades of brown in their plumage and show little difference between the sexes. They tend to forage on or near the ground; their stout, conical bills are used for eating seeds.

TYPICAL SPARROW
A plump White-crowned Sparrow shows the typical stout emberizid beak.

Tanagers

Tanagers feature brightly colored males and dull, olive-colored females. Relatively sluggish birds, they feed mostly on insects like bees and wasps, and fruit. They have similar songs but distinctive calls.

MALE COLORS
Male Western Tanagers are some of North America's most colorful birds.

Cardinals & Buntings

Males of this family are among the showiest birds in Canada, both physically and vocally. They range in color from the red of the Northern Cardinal to the electric-blue Indigo Bunting.

STRONG BILLS
Male Northern Cardinals have impressive bills, perfect for cutting open seed hulls.

McCown's Longspur

Rhynchophanes mccownii

Male McCown's Longspurs can often be found performing their flight displays on wings held in a V position. A dull female resembles a female House Sparrow, but can be distinguished by the white patches on its tail. Recent evidence suggests that McCown's Longspur may be more closely related to the Snow Bunting than to the other species of longspurs.

♂ BREEDING

black "T" on white tail

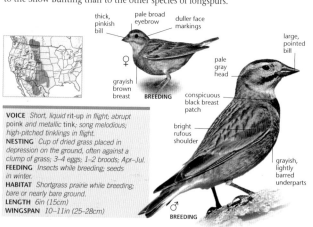

thick, pinkish bill

pale broad eyebrow

duller face markings

♀

grayish brown breast

BREEDING

large, pointed bill

pale gray head

conspicuous black breast patch

bright rufous shoulder

grayish, lightly barred underparts

♂ BREEDING

VOICE *Short, liquid rit-up in flight; abrupt poink and metallic tink; song melodious; high-pitched tinklings in flight.*
NESTING *Cup of dried grass placed in depression on the ground, often against a clump of grass; 3–4 eggs; 1–2 broods; Apr–Jul.*
FEEDING *Insects while breeding; seeds in winter.*
HABITAT *Shortgrass prairie while breeding; bare or nearly bare ground.*
LENGTH *6in (15cm)*
WINGSPAN *10–11in (25–28cm)*

Lapland Longspur

Calcarius lapponicus

One of the most numerous breeding birds of the Arctic tundra, the Lapland Longspur is found in huge flocks over open habitats in their winter range. They can be seen on gravel roads and in barren countryside immediately following heavy snowfalls. Genetic evidence suggests that the four longspur species and the two *Plectrophenax* buntings may form a distinct group of their own.

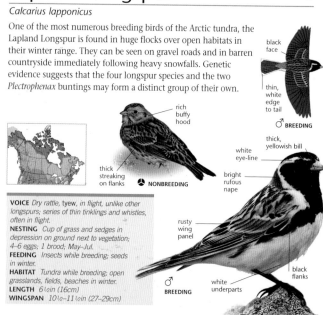

black face

thin, white edge to tail

♂ BREEDING

rich buffy hood

thick streaking on flanks

🌑 NONBREEDING

thick, yellowish bill

white eye-line

bright rufous nape

rusty wing panel

white underparts

♂ BREEDING

black flanks

VOICE *Dry rattle, tyew, in flight, unlike other longspurs; series of thin tinklings and whistles, often in flight.*
NESTING *Cup of grass and sedges in depression on ground next to vegetation; 4–6 eggs; 1 brood; May–Jul.*
FEEDING *Insects while breeding; seeds in winter.*
HABITAT *Tundra while breeding; open grasslands, fields, beaches in winter.*
LENGTH *6½in (16cm)*
WINGSPAN *10½–11½in (27–29cm)*

Smith's Longspur (S)

Calcarius pictus

On both its remote Arctic breeding grounds and its restricted shortgrass range in winter, Smith's Longspur hides on the ground, making it very hard to spot. It migrates through the Great Plains in fall, but on the return journey it swings east, giving it an elliptical migration path. Both males and females mate with several partners.

white cheek patch

white outer tail feathers

♂ **BREEDING**

relatively long wings

rich, buffy overall

fine breast streaks

wings extend past tail

♀

black-and-white "helmet"

thin bill

orange collar

white shoulder

white undertail feathers

rich pumpkin-colored underparts

♂ **BREEDING**

VOICE *Mechanical, dry, sharp rattle in flight; nasal* nief *when squabbling; series of thin, sweet whistles.*
NESTING *Concealed cup of sedges, lined with feathers, in hummock on ground; 3–5 eggs; 1 brood; Jun–Jul.*
FEEDING *Seeds and insects; introduced foxtail grass while migrating.*
HABITAT *Tundra-taiga timberline while breeding; shortgrass prairie while migrating.*
LENGTH *6–6½in (15–16cm)*
WINGSPAN *10–11½in (25–29cm)*

Chestnut-collared Longspur (T)

Calcarius ornatus

The Chestnut-collared Longspur traditionally bred in areas of the western prairies that had been recently disturbed by roaming herds of bison or wild fires. This habitat is now hard to find, and populations have declined. It can be distinguished by the triangular black patch on its tail and the breeding male's black belly.

white patch on wing

♂ **BREEDING**

buff eyebrow

gray-brown overall

♀ **NONBREEDING**

white eyebrow

chestnut neck

streaked upperparts

tan cheeks

black underparts

♂ **BREEDING**

white outer tail feathers

VOICE *Chortling* KTI-uhl-uh *in flight, often in series; soft rattle and short buzz.*
NESTING *Grassy cup on ground, in grass clump or next to rock; 3–5 eggs; 1–2 broods; May–Aug.*
FEEDING *Seeds year-round; also insects when breeding.*
HABITAT *Shortgrass prairie while breeding; grasslands and cultivated fields during migration.*
LENGTH *5½–6in (14–15cm)*
WINGSPAN *10–10½in (25–27cm)*

Snow Bunting

Plectrophenax nivalis

The bold white wing patches of the Snow Bunting make it recognizable in a winter flock of dark-winged longspurs and larks. To secure the best territories, some male Snow Buntings arrive as early as April in their barren high-Arctic breeding grounds. The species is very similar in appearance to the rare and localized McKay's Bunting, which generally has less black on the back, wings, and tail, and the two species can be difficult to discern.

less white in wings

white outer tail feathers

♂ NONBREEDING

yellow bill

large white patches on black wings

♀ NONBREEDING

gray body

white eye-ring

rusty orange cheek patch

dark brown eyes

black peeks through buffy feather edgings

rusty orange breast patch

♂ NONBREEDING

white underparts

VOICE Musical, liquid rattle in flight, also *tyew* notes and short buzz; pleasant series of squeaky and whistled notes.
NESTING Bulky cup of grass and moss, lined with feathers, in sheltered rock crevice; 3–6 eggs; 1 brood; Jun–Aug.
FEEDING Seeds (sedge in Arctic), flies and other insects, and buds on migration.
HABITAT Rocky areas, usually near sparsely vegetated tundra, while breeding; open areas and shores in winter.
LENGTH 6½–7in (16–18cm)
WINGSPAN 12½–14in (32–35cm)

Lark Bunting ⓢ

Calamospiza melanocorys

The stocky Lark Bunting—unlike the Chestnut-collared Longspur—has coped well with the changes to its habitat by humans, occurring in extraordinary density throughout its range. Nomadic flocks of thousands scour the high deserts, open grasslands, and sage bush for seeds. Breeding males are black with large white wing patches; females and immature birds are duller.

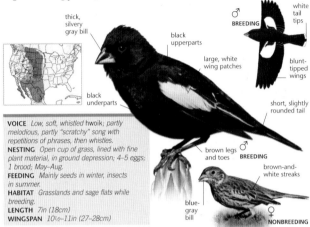

thick, silvery gray bill

black upperparts

large, white wing patches

black underparts

♂ BREEDING

white tail tips

blunt-tipped wings

short, slightly rounded tail

brown legs and toes — ♂ BREEDING

brown-and-white streaks

blue-gray bill

♀ NONBREEDING

VOICE *Low, soft, whistled hwoik; partly melodious, partly "scratchy" song with repetitions of phrases, then whistles.*
NESTING *Open cup of grass, lined with fine plant material, in ground depression; 4–5 eggs; 1 brood; May–Aug.*
FEEDING *Mainly seeds in winter, insects in summer.*
HABITAT *Grasslands and sage flats while breeding.*
LENGTH *7in (18cm)*
WINGSPAN *10½–11in (27–28cm)*

Fox Sparrow ⓢ

Passerella iliaca

Larger and more colorful than its close relatives, the Fox Sparrow forages by crouching low in leaf litter and hopping about to disturb leaves, under which it finds seeds or insects. It varies considerably over its huge range, from thick-billed birds in the Sierras to dark ones in the Northwest, and distinctive reds in the East.

dark rufous overall

🔴 RED

gray head and back

P. i. altivagans
SLATE-COLORED

rusty streaks on back

two white wing bars

gray nape

darker upper mandible

🔴 RED

belly marked with rufous chevrons

long, rusty tail

VOICE *Sharp, dry tshak or tshuk; high-pitched tzeep! in flight; complex, musical song with trills and whistles.*
NESTING *Dense cup of grasses or moss lined with fine material; usually placed low in shrub; 2–5 eggs; 1 brood; Apr–Jul.*
FEEDING *Forages for insects, seeds, and fruit.*
HABITAT *Boreal forest zone; coastal or near-coastal thickets in coniferous forests or mixed woodlands.*
LENGTH *6–7½in (15–19cm)*
WINGSPAN *10½–11½in (27–29cm)*

Song Sparrow

S

Melospiza melodia

The familiar song of this species can be heard in backyards across the continent, including in winter, although it varies both individually and geographically. The Song Sparrow may be the North American champion of geographic variation: about 30 subspecies have been described; most have streaky plumage and can be found perched in low shrubs in open, shrubby, or wet areas. Males sing in spring and summer on an exposed perch around eye level.

grayish head with dark chestnut brown crown

heavily streaked brownish gray upperparts

WEST COAST

streaked underparts

dark "mustache" bordering whitish throat

heavily streaked underparts

long, dark, rounded tail

whitish lower belly

paler neck

more rusty overall

M. m. saltonis **SOUTHWEST**

grayish head with brown markings

central breast spot

M. m. melodia **EASTERN**

VOICE *Dry* tchip; *clear* siiiti *in flight; jumble of variable whistles and trills,* deeeep deeeep deep-deep chrrrr tiiiiiiiiiiiii tyeeur.
NESTING *Bulky cup on or near ground, in brush or marsh vegetation; 3–5 eggs; 1–3 broods; Mar–Aug.*
FEEDING *Mainly insects in summer; mainly seeds, but also fruit in winter.*
HABITAT *Variety of open habitats, including fields and pastures, the edges of forests and wetlands, and suburbs.*
LENGTH *5–7½in (13–19cm)*
WINGSPAN *8½–12in (21–31cm)*

Lincoln's Sparrow Ⓢ

Melospiza lincolnii

In the breeding season, Lincoln's Sparrow seeks out predominantly moist willow scrub at the tundra–taiga timberline. It may visit backyard feeders in winter, but it generally prefers to stay within fairly dense cover. Its rich, musical song is unmistakable and varies remarkably little from region to region.

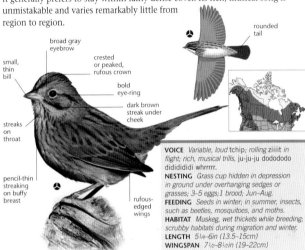

rounded tail

broad gray eyebrow

crested or peaked, rufous crown

small, thin bill

bold eye-ring

dark brown streak under cheek

streaks on throat

pencil-thin streaking on buffy breast

rufous-edged wings

VOICE *Variable, loud* tchip; *rolling* ziiiit *in flight; rich, musical trills, ju-ju-ju dodododo didididi whrrrrr.*
NESTING *Grass cup hidden in depression in ground under overhanging sedges or grasses; 3–5 eggs; 1 brood; Jun–Aug.*
FEEDING *Seeds in winter; in summer, insects, such as beetles, mosquitoes, and moths.*
HABITAT *Muskeg, wet thickets while breeding; scrubby habitats during migration and winter.*
LENGTH 5¼–6in (13.5–15cm)
WINGSPAN 7½–8½in (19–22cm)

Swamp Sparrow Ⓢ

Melospiza georgiana

The Swamp Sparrow is a common breeder in wet habitats, especially tall reed marshes. It is skittish and often seen darting rapidly into cover. Though often confused with the Song and Lincoln's Sparrow, the Swamp Sparrow has very faint, blurry streaking on its gray breast, and sports conspicuous rusty-edged wing feathers.

gray and rufous face

rufous crown

rufous flanks

tan upperparts with dark streaks

BREEDING

gray breast with fine streaking

dark, rounded tail

BREEDING

rusty margins to wing feathers

unstreaked gray nape

tawny flanks

NONBREEDING

VOICE *Slightly nasal, forceful* chimp; *high, buzzy* ziiiii *in flight; slow, monotonous, loose trill of chirps.*
NESTING *Bulky cup of dry plants placed above water in marsh vegetation; 3–5 eggs; 1–2 broods; May–Jul.*
FEEDING *Mostly insects while breeding, especially grasshoppers; seeds in winter.*
HABITAT *Marshes, cedar bogs, damp meadows, and wet hayfields.*
LENGTH 5–6in (12.5–15cm)
WINGSPAN 7–7½in (18–19cm)

Harris's Sparrow

(S)

Zonotrichia querula

An unmistakable black-faced, pink-billed bird, Harris's Sparrow is the only breeding bird endemic to Canada. It can be seen in the US during migration or in winter on the Great Plains and is occasionally found among large flocks of White-throated and White-crowned Sparrows. It is a large sparrow, approaching the Northern Cardinal in size.

indistinct facial markings **NONBREEDING**

pinkish bill

two wing bars

NONBREEDING

black crown

gray cheeks

pinkish or yellow bill

black cheek patch

gray rump and undertail feathers

black chin and throat

BREEDING

VOICE *Sharp* weeek; *melancholy series of 2–4 whistles on the same pitch.*
NESTING *Bulky cup placed on ground among vegetation or near ground in brush; 3–5 eggs; 1 brood; Jun–Aug.*
FEEDING *Seeds, insects, buds, and even young conifer needles in summer.*
HABITAT *Scrub-tundra along taiga-tundra timberline while breeding; winters in prairies.*
LENGTH *6¾–7½in (17–19cm)*
WINGSPAN *10½–11in (27–28cm)*

White-crowned Sparrow

(S)

Zonotrichia leucophrys

The White-crowned Sparrow has four subspecies. Pacific Coast birds are brown, with a yellowish bill and a gray-washed head stripe; western and northwestern birds are gray below, with an orange bill and a white head stripe; Eastern and Rocky Mountain birds have a pink bill and a bright white head stripe; birds in southwest Canada are darker.

gray rump and uppertail

longish tail

brown crown

two wing bars

white crown with two black stripes

yellowish bill

gray breast

black line

gray cheek

two wing bars

white streaking on brown upperparts

unmarked, grayish underparts

Z. l. oriantha
INTERIOR WEST

VOICE *Sharp tink; thin seep in flight; buzzy whistle followed by buzzes, trills, and whistles.*
NESTING *Bulky cup of grass placed on or near the ground in bushes; 4–6 eggs; 1–3 broods; Mar–Aug.*
FEEDING *Seeds, insects, fruit, buds, and grass.*
HABITAT *Boreal forest and tundra limit; breeds in willow thickets, wet forest, and suburbs.*
LENGTH *6½–7in (16–18cm)*
WINGSPAN *9½–10in (24–26cm)*

White-throated Sparrow

Zonotrichia albicollis

White-throated Sparrows sing all year round; its whistled, rhythmic song can be remembered with the mnemonic *Oh sweet Canada Canada Canada*. This species has two different color forms, one with a white stripe above its eye, and one with a tan stripe. In the nonbreeding season, large flocks roam the leaf litter of woodlands in search of food.

two white wing bars

tan stripe

browner face

bold white stripe

yellow patch

bright rufous back and tail

white throat

TAN-STRIPED

gray underparts

WHITE-STRIPED

fairly long tail

VOICE Loud, sharp jink; lisping tssssst! in flight; clear whistle comprising 1–2 higher notes, then three triplets.
NESTING Cup placed on or near ground in dense shrubbery; 2–6 eggs; 1 brood; May–Aug.
FEEDING Seeds, fruit, insects, buds, and various grasses.
HABITAT Forests while breeding; wooded thickets and hedges.
LENGTH 6½–7½in (16–17.5cm)
WINGSPAN 9–10in (23–26cm)

Golden-crowned Sparrow

Zonotrichia atricapilla

The Golden-crowned Sparrow's minor-key song has been likened to *I'm so tired* or *No gold here*. It has been regarded as a pest in the past because of its habit of consuming crops in agricultural fields and gardens. Nonbreeding adults retain their distinctive golden crown in the winter, but it appears duller.

white wing bars

BREEDING

duller yellow on crown

much less black on face

bright yellow crown

thick black eyebrow

NONBREEDING

light grayish brown underparts

long tail

BREEDING

VOICE Loud tsik; soft, short seep in flight; series of melancholy whistles, sometimes slurred or trilled.
NESTING Concealed bulky cup placed on ground at base of bush; 3–5 eggs; 1–2 broods; Jun–Aug.
FEEDING Seeds, insects, fruit, flowers, buds.
HABITAT Shrubby habitat along the treeline and open, boggy forests while breeding; dense thickets in winter.
LENGTH 7in (18cm)
WINGSPAN 9–10in (23–25cm)

Dark-eyed Junco

S

Junco hyemalis

The Dark-eyed Junco's appearance at birdfeeders during snowstorms has earned it the colloquial name of "snowbird." Sixteen subspecies have been described. "Slate-colored" populations are widespread across Canada and the northeastern US, "Pink-sided" birds breed in Idaho, Montana, and Wyoming, and "Oregon" birds breed in the Pacific West, from Alaska to British Columbia and the mountainous western US in the Sierras south to Mexico.

white outer tail feathers

♂ SLATE-COLORED

dark area between eye and bill

bluish gray hood

dull, brownish back

♀ PINK-SIDED

pinkish flanks

blackish hood

rust back

reddish flanks

♂ OREGON

gray body with brown wash to back

dark gray head

white belly

♂ SLATE-COLORED

VOICE *Loud, smacking* tick *and soft* dyew; *rapid, twittering, and buzzy* zzeet *in flight; simple, liquid, 1-pitch trill.*
NESTING *Cup placed on ground hidden under vegetation or next to rocks; 3–5 eggs; 1–2 broods; May–Aug.*
FEEDING *Insects and seeds; also berries; forages on ground at base of trees and shrubs or under feeders.*
HABITAT *Coniferous and mixed forests while breeding; open woodlands, fields, parks, and backyards in winter.*
LENGTH 6–6¾in (15–17cm)
WINGSPAN 8–10in (20–26cm)

Savannah Sparrow

Passerculus sandwichensis

The Savannah Sparrow shows tremendous variation—21 subspecies— across its vast range, but it is always brown, with dark streaks above and white with dark streaks below. The pale "Ipswich Sparrow" (*P. s. princeps*) breeds on Sable Island, Nova Scotia, and winters along the East Coast.

brown overall

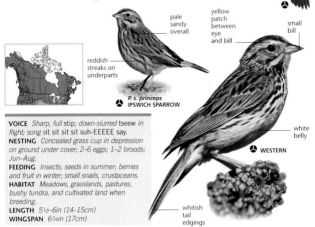

pale sandy overall

yellow patch between eye and bill

small bill

reddish streaks on underparts

P. s. princeps
IPSWICH SPARROW

white belly

WESTERN

whitish tail edgings

VOICE *Sharp, full stip; down-slurred tseew in flight; song sit sit sit sit suh-EEEEE say.*
NESTING *Concealed grass cup in depression on ground under cover; 2–6 eggs; 1–2 broods; Jun–Aug.*
FEEDING *Insects; seeds in summer; berries and fruit in winter; small snails, crustaceans.*
HABITAT *Meadows, grasslands, pastures, bushy tundra, and cultivated land when breeding.*
LENGTH 5½–6in (14–15cm)
WINGSPAN 6¾in (17cm)

Nelson's Sharp-tailed Sparrow

Ammodramus nelsoni

This sparrow has three subspecies that differ in plumage, breeding habitat, and location. *A. n. nelsoni* is the most brightly colored and is found from the southern Northwest Territories south to northwest Wisconsin. The duller *A. n. subvirgatus* breeds in coastal Maine and the Maritimes, and along the St. Lawrence River. The intermediate-looking *A. n. alterus* breeds along the coasts of Hudson Bay.

dark, rounded, spiky tail

no bold streaks on underparts

A. n. subvirgatus

smaller bill

brighter upperparts

A. n. nelsoni

dark cheek marks

streaked, washed-out pattern on back

medium-sized bill

short, pointed tail

faint streaking on underparts

A. n. subvirgatus

VOICE *Sharp tik; husky t-SHHHHEE-uhrr.*
NESTING *Open cup of grass, sometimes built up on sides to curve inwards, placed on or just above ground; 4–5 eggs; 1 brood; May–Jul.*
FEEDING *Forages on the ground among insects, spiders, and seeds, in dense grass or at edges of pools.*
HABITAT *Freshwater or saltwater marsh habitats while breeding; nonbreeders favor wet weedy fields.*
LENGTH 4¾in (12cm)
WINGSPAN 7in (17.5cm)

Le Conte's Sparrow

S

Ammodramus leconteii

The tiny Le Conte's Sparrow is one of the smallest of all sparrows and is usually difficult to see. It prefers to dart for cover under grasses instead of flushing when disturbed, and its flight call and song are remarkably insect-like. Many people who hear it often pass off the unseen bird as a grasshopper.

boldly striped back

pale, tawny rump

spiky tail

gray ear patch

white median crown stripe

orange eyebrow

small bill

purplish and gray streaks on nape

orange throat

white-edged wing feathers

fine streaks on buffy breast

VOICE *Long, down-slurred zheeep; flight call similar to grasshopper; insect-like, buzzy tik'-uht-tizz-ZHEEEEEE-k.*
NESTING *Concealed little cup placed on or near ground; 3–5 eggs; 1 brood; Jun–Aug.*
FEEDING *Forages on ground in grasses for insects, insect larvae, spiders, seeds.*
HABITAT *Marshes, wet meadows, and bogs while breeding; tall grasses and marshes in winter.*
LENGTH 4½–5in (11.5–13cm)
WINGSPAN 6½–7in (16–18cm)

Baird's Sparrow

D

Ammodramus bairdii

The sweet, tinkling song of Baird's Sparrow is quite different to the buzzy songs of the other *Ammodramus* sparrows. Its square, pale-edged tail is also unique within its genus. It is usually seen only on its breeding grounds and is difficult to spot elsewhere, scurrying out of sight if disturbed.

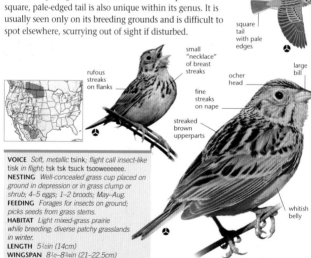

streaked upperparts

square tail with pale edges

small "necklace" of breast streaks

rufous streaks on flanks

ocher head

large bill

fine streaks on nape

streaked brown upperparts

whitish belly

VOICE *Soft, metallic tsink; flight call insect-like tisk in flight; tsk tsk tsuck tsooweeeeee.*
NESTING *Well-concealed grass cup placed on ground in depression or in grass clump or shrub; 4–5 eggs; 1–2 broods; May–Aug.*
FEEDING *Forages for insects on ground; picks seeds from grass stems.*
HABITAT *Light mixed-grass prairie while breeding; diverse patchy grasslands in winter.*
LENGTH 5½in (14cm)
WINGSPAN 8½–8¾in (21–22.5cm)

Henslow's Sparrow

Ammodramus henslowii

The combination of a large, flat, greenish head, and purplish back are unique to Henslow's Sparrow. While it has suffered greatly from the drainage, cultivation, and urbanization of much of its preferred breeding grounds, the Henslow's Sparrow has also recently started to use reclaimed strip mines in northwest Missouri and Iowa for breeding.

dark reddish overall

round, spiky tail

flat, greenish head with black stripes

heavy bill

whitish scaling on purplish back

rufous-edged wing feathers

black streaks on buffy breast

VOICE Sharp tsik; long, high, shrill tseeeeee in flight; hiccupping sputter with second note higher tsih-LIK!
NESTING Cup of loosely woven grass placed on or near ground; 2–5 eggs; 1–2 broods; May–Aug.
FEEDING Seeds; insects, insect larvae, and spiders in summer.
HABITAT Tallgrass prairie and wet grasslands while breeding.
LENGTH 4¾–5in (12–13cm)
WINGSPAN 6½in (16cm)

Grasshopper Sparrow

Ammodramus savannarum

Although its large head and spiky tail are typical of its genus, the Grasshopper Sparrow is the only *Ammodramus* sparrow to have a plain breast and two completely different songs. Its common name derives from its song, which resembles the sounds grasshoppers make. It varies geographically, with about 12 subspecies.

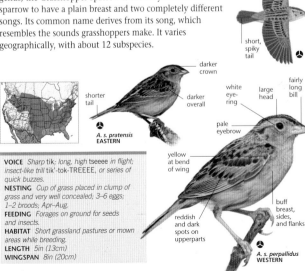

buff overall

short, spiky tail

darker crown

shorter tail

darker overall

A. s. pratensis
EASTERN

white eye-ring

large head

fairly long bill

pale eyebrow

yellow at bend of wing

reddish and dark spots on upperparts

buff breast, sides, and flanks

A. s. perpallidus
WESTERN

VOICE Sharp tik; long, high tseeee in flight; insect-like trill tik'-tok-TREEEE, or series of quick buzzes.
NESTING Cup of grass placed in clump of grass and very well concealed; 3–6 eggs; 1–2 broods; Apr.–Aug.
FEEDING Forages on ground for seeds and insects.
HABITAT Short grassland pastures or mown areas while breeding.
LENGTH 5in (13cm)
WINGSPAN 8in (20cm)

Chipping Sparrow

Spizella passerina

The Chipping Sparrow is a common, trusting bird, which breeds in backyards across most of North America. While they are easily identifiable in the summer, "Chippers" molt into a drab, nonbreeding plumage during fall, at which point they are easily confused with the Clay-colored and Brewer's Sparrows they flock with. Most reports of this species across the north in winter are actually of the larger American Tree Sparrow.

pale
underparts

rusty
cast to
crown

pinkish
bill

bright
rufous
crown

blackish
bill

white
eyebrow

black
eye
line

heavily
streaked,
especially
on breast

gray
underparts

cleft
tail

BREEDING

VOICE *Sharp* tsip; *sharp, thin* tsiiit *in flight; insect-like trill of* chip *notes, variable in duration.*
NESTING *Nest cup usually placed well off the ground in tree or shrub; 3–5 eggs; 1–2 broods; Apr–Aug.*
FEEDING *Seeds of grasses and annuals, some fruits; insects and other invertebrates while breeding.*
HABITAT *Open forest, woodlands, grassy park-like areas, shorelines, and backyards.*
LENGTH *5½in (14cm)*
WINGSPAN *8½in (21cm)*

American Tree Sparrow

Spizella arborea

Commonly mistaken for the smaller Chipping Sparrow, the American Tree Sparrow's central breast spot, bicolored bill, and large size are unique among the *Spizella*. As a highly social and vocal species, large, noisy winter flocks can be found feeding in weedy fields and along the roadsides of southern Canada and the northern US.

BREEDING

striped back

cleft tail

NONBREEDING

gray head and nape

rufous crown

black-and-yellow bill

rusty stripe behind eye

rust patch on shoulder

black and rust streaking on back

dark, central spot

long, squarish tail

BREEDING

VOICE *Bell-like* teedle-ee; *thin, slightly descending* tsiiiu *in flight;* seee seee di-di-di di-di-di dyew dyew.
NESTING *Neat cup on ground concealed within thicket; 4–6 eggs; 1 brood; Jun–Jul.*
FEEDING *Seeds, berries, and variety of insects.*
HABITAT *Scrubby thickets of birch and willows while breeding; open brushy habitats for nonbreeders.*
LENGTH *6¼in (16cm)*
WINGSPAN *9½in (24cm)*

Field Sparrow

Spizella pusilla

The distinctive trill of the Field Sparrow's song is a characteristic sound of scrubby areas in southeastern Canada. Its bright-pink bill, plain "baby face," and white eye-ring make this sparrow one of the easiest to identify. It has brighter plumage in the East, and drabber plumage in the interior part of its range.

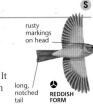

rusty markings on head

long, notched tail

REDDISH FORM

white eye-ring

streaking on back

white wing bars

GRAYISH FORM

small, pink bill

light rust cheek and crown

tan underparts

long tail

distinctive pink legs

REDDISH FORM

VOICE *Sharp* tsik; *strongly descending* tsiiiu *in flight; series of sweet, down-slurred whistles accelerating to a rapid trill.*
NESTING *Grass cup placed on or just above ground in grass or bush; 3–5 eggs; 1–3 broods; Mar–Aug.*
FEEDING *Seeds; insects, insect larvae, and spiders in summer.*
HABITAT *Overgrown fields, woodland edges, and roadsides while breeding.*
LENGTH *5½in (14cm)*
WINGSPAN *8in (20cm)*

Clay-colored Sparrow

Spizella pallida

The little Clay-colored Sparrow is best known for its mechanical, buzzy song. It spends much of its foraging time away from the breeding habitat; consequently, males' territories are quite small, allowing for dense breeding populations. During the nonbreeding season, they form large flocks in open country, associating with other *Spizella* sparrows, especially Chipping and Brewer's.

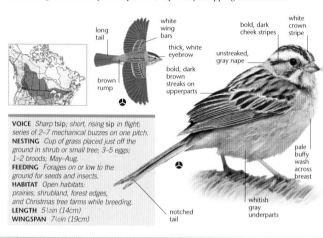

long tail

white wing bars

thick, white eyebrow

brown rump

bold, dark brown streaks on upperparts

bold, dark cheek stripes

unstreaked, gray nape

white crown stripe

pale buffy wash across breast

whitish gray underparts

notched tail

VOICE Sharp tsip; short, rising sip in flight; series of 2–7 mechanical buzzes on one pitch.
NESTING Cup of grass placed just off the ground in shrub or small tree; 3–5 eggs; 1–2 broods; May–Aug.
FEEDING Forages on or low to the ground for seeds and insects.
HABITAT Open habitats: prairies, shrubland, forest edges, and Christmas tree farms while breeding.
LENGTH 5½in (14cm)
WINGSPAN 7½in (19cm)

Brewer's Sparrow

Spizella breweri

This sparrow's conspicuous eye-ring and streaked nape are good identification features, along with its varied, loud, trilling and chattering song. Most Brewer's Sparrows nest in the western US, but an isolated population, subspecies "Timberline," breeds in the Canadian Rockies and into Alaska. It is usually darker, more boldly marked, and longer-billed with a lower, slower, more musical song than its relative.

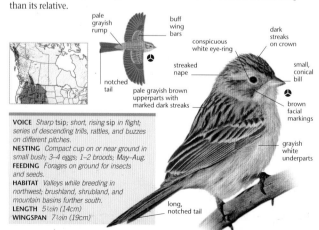

pale grayish rump

buff wing bars

conspicuous white eye-ring

notched tail

streaked nape

pale grayish brown upperparts with marked dark streaks

dark streaks on crown

small, conical bill

brown facial markings

grayish white underparts

long, notched tail

VOICE Sharp tsip; short, rising sip in flight; series of descending trills, rattles, and buzzes on different pitches.
NESTING Compact cup on or near ground in small bush; 3–4 eggs; 1–2 broods; May–Aug.
FEEDING Forages on ground for insects and seeds.
HABITAT Valleys while breeding in northwest; brushland, shrubland, and mountain basins further south.
LENGTH 5½in (14cm)
WINGSPAN 7½in (19cm)

Vesper Sparrow

Pooecetes gramineus

The Vesper Sparrow is named for its sweet evening song. The Vesper Sparrow needs areas with bare ground to breed, so it is one of the few species that can successfully nest in areas of intense agriculture. Despite that, its numbers are declining to the point of being endangered in some regions of Canada.

rusty shoulders

uniformly colored and streaked overall

bold white eye-ring

dark-bordered ear patches

white outer tail feathers

pale brown upperparts

streaked breast

bold white-edged long, dark, square tail

VOICE *Full* tchup; *thin* tseent *in flight; 2 whistles of same pitch, followed by 2 higher-pitched ones, then trills, ends lazily.*
NESTING *Cup placed on patch of bare ground, against grass, bush, or rock; 3–5 eggs; 1 brood; Apr–Aug.*
FEEDING *Insects and seeds.*
HABITAT *Sparse grassland, cultivated fields, recently burned areas, and mountain parks while breeding.*
LENGTH *6¼in (16cm)*
WINGSPAN *10in (25cm)*

Lark Sparrow

Chondestes grammacus

The bold harlequin face pattern, single central breast spot, and long, rounded, black tail with white corners make the Lark Sparrow one of the most easily identifiable of all sparrows. It is commonly found singing from the top of a fencepost or small tree in southern Canadian prairies. Male birds are strongly territorial of their nesting sites.

rounded tail with white corners

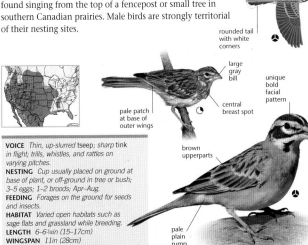

large gray bill

unique bold facial pattern

pale patch at base of outer wings

central breast spot

brown upperparts

pale plain rump

VOICE *Thin, up-slurred* tseep; *sharp* tink *in flight; trills, whistles, and rattles on varying pitches.*
NESTING *Cup usually placed on ground at base of plant, or off-ground in tree or bush; 3–5 eggs; 1–2 broods; Apr–Aug.*
FEEDING *Forages on the ground for seeds and insects.*
HABITAT *Varied open habitats such as sage flats and grassland while breeding.*
LENGTH *6–6¾in (15–17cm)*
WINGSPAN *11in (28cm)*

Spotted Towhee

Pipilo maculatus

This large, colorful sparrow can often be heard rummaging through dry leaves in the undergrowth in search of food, using its feet like a garden rake. The Spotted Towhee has been separated into 20 rather complex subspecies, but all are distinguished from the Eastern Towhee by the presence of white spots and bars on their upperwings.

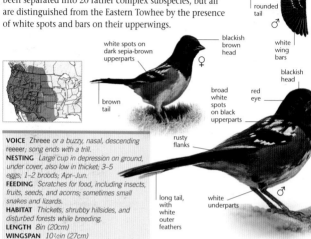

white tail tips

rounded tail ♂

white wing bars

white spots on dark sepia-brown upperparts

blackish brown head ♀

brown tail

blackish head

broad white spots on black upperparts

red eye

rusty flanks

long tail, with white outer feathers

white underparts ♂

VOICE Zhreee *or a buzzy, nasal, descending* reeeer; *song ends with a trill.*
NESTING *Large cup in depression on ground, under cover, also low in thicket; 3–5 eggs; 1–2 broods; Apr–Jun.*
FEEDING *Scratches for food, including insects, fruits, seeds, and acorns; sometimes small snakes and lizards.*
HABITAT *Thickets, shrubby hillsides, and disturbed forests while breeding.*
LENGTH *8in (20cm)*
WINGSPAN *10½in (27cm)*

Eastern Towhee

Pipilo erythrophthalmus

The Eastern Towhee is famous for its vocalizations and has one of the best-known mnemonics for its song: "drink your tea." Like all towhees, the Eastern Towhee feeds noisily by jumping backwards with both feet at once to move leaves and reveal the insects and seeds hidden underneath.

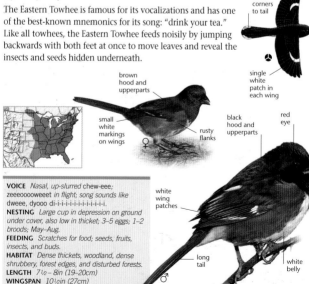

white corners to tail

single white patch in each wing

brown hood and upperparts

small white markings on wings ♀

rusty flanks

black hood and upperparts

red eye

white wing patches

long tail ♂

white belly

VOICE *Nasal, up-slurred* chew-eee; zeeeooooweeet *in flight; song sounds like* dwee, dyooo *di-i-i-i-i-i-i-i-i-i-i.*
NESTING *Large cup in depression on ground under cover, also low in thicket; 3–5 eggs; 1–2 broods; May–Aug.*
FEEDING *Scratches for food; seeds, fruits, insects, and buds.*
HABITAT *Dense thickets, woodland, dense shrubbery, forest edges, and disturbed forests.*
LENGTH *7½ – 8in (19–20cm)*
WINGSPAN *10½in (27cm)*

Scarlet Tanager

S

Piranga olivacea

Although the male breeding Scarlet Tanager is one of the brightest and most easily identified North American birds, its secretive nature and preference for the canopies of well-shaded oak woodlands makes it difficult to spot. Males can vary in appearance—some are orange, not scarlet, and others have a faint reddish wing bar.

black wings

red body

tail appears short in flight

♂ BREEDING

greenish rump and upper tail

♀

overall greenish upperparts

dark brown eyes

vibrant scarlet head and body

grayish yellow bill

black wings

black tail

dark gray feet and legs

♂ BREEDING

VOICE *Hoarse, drawn out CHIK-breeer or CHIK; upslurred, whistled* pwee *in flight; burry, slurred querit-queer-query-querit-queer.*
NESTING *Loosely woven grass cup, lined with fine material, high in tree; 3–5 eggs; 1 brood; May–Jul.*
FEEDING *Insects, larvae, fruit, buds, and berries.*
HABITAT *Mature deciduous and mixed forests, especially with large oaks, while breeding.*
LENGTH *7in (18cm)*
WINGSPAN *11½in (29cm)*

Western Tanager

S

Piranga ludoviciana

The hoarse song of the male Western Tanager is a characteristic sound of coniferous forests in western North America. All *Piranga* tanagers have songs and whistled flight calls that closely resemble those of the *Pheucticus* grosbeaks. Recent genetic studies suggest that *Piranga* tanagers are actually part of the *Pheucticus* grosbeak family.

orange head

two wing bars

♂

BREEDING

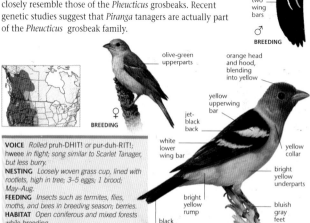

olive-green upperparts

♀
BREEDING

orange head and hood, blending into yellow

yellow upperwing bar

jet-black back

white lower wing bar

yellow collar

bright yellow underparts

bright yellow rump

bluish gray feet and legs

black tail

♂ BREEDING

VOICE *Rolled pruh-DHIT! or pur-duh-RIT!; hweee in flight; song similar to Scarlet Tanager, but less burry.*
NESTING *Loosely woven grass cup, lined with rootlets, high in tree; 3–5 eggs; 1 brood; May–Aug.*
FEEDING *Insects such as termites, flies, moths, and bees in breeding season; berries.*
HABITAT *Open coniferous and mixed forests while breeding.*
LENGTH *7½in (19cm)*
WINGSPAN *11½in (29cm)*

Dickcissel

Spiza americana

The Dickcissel is a tallgrass prairie specialist and seldom breeds outside this core range. Immature birds, without yellow and rusty plumage, are very similar to female House Sparrows—vagrant and wintering Dickcissels in North America are often mistaken for sparrows. It winters in Venezuela, where it is a notorious pest to seed crops.

streaked back

♂ BREEDING

yellow-tinged, long eye-line

bold braces on back

large, pointed bill

yellow eyebrow

gray nape

black "v" on yellow breast

rufous shoulder

finely streaked underparts

♀

VOICE *Flat chik; low, electric buzz frrrrrrrt in flight; insect-like stutters followed by longer chirps or trill* dick-dick-dick-SISS-SISS-suhl.
NESTING *Bulky cup near ground in dense vegetation; 3–6 eggs; 1–2 broods; May–Aug.*
FEEDING *Insects, spiders, and seeds from ground.*
HABITAT *Tallgrass prairie, grassland, hayfields, unmown roadsides, and untilled cropfields while breeding.*
LENGTH 6½in (16cm)
WINGSPAN 9½in (24cm)

♂ BREEDING

Black-headed Grosbeak

Pheucticus melanocephalus

This orange-breasted grosbeak is the western counterpart of the Rose-breasted Grosbeak. The two species are closely related, despite their color differences, and interbreed where their ranges meet in the Great Plains. The Black-headed Grosbeak is aggressive on its breeding grounds, with both sexes fighting off intruders.

♂ BREEDING

large, white wing patch

white eyebrow

brown head and upperparts

tawny breast

fine brown streaks on underparts

♀

thick, gray-and-black bill

black head

orange hind collar

black upperparts

two white wing bars

orange rump

♂

orange breast

short tail, with white corners

VOICE *Hwik, similar to Rose-breasted Grosbeak, but flatter and less squeaky; song generally higher, faster, less fluid, and harsher.*
NESTING *Loose, open cup or platform, usually in deciduous sapling, just above eye level; 3–5 eggs; 1–2 broods; May–Sep.*
FEEDING *Insects, spiders; also seeds, fruit.*
HABITAT *Dense deciduous growth, such as old fields, hedgerows, and disturbed forests, while breeding.*
LENGTH 8½in (21cm)
WINGSPAN 12½in (32cm)

Rose-breasted Grosbeak

Pheucticus ludovicianus

For many birdwatchers in the East, the appearance of a flock of dazzling male Rose-breasted Grosbeaks in early May signals the peak of spring songbird migration. Females and immature males have more somber plumage. In the fall, immature male Rose-breasted Grosbeaks often have orange breasts and are commonly mistaken for female Black-headed Grosbeaks. The difference is in the pink wing lining usually visible on perched birds, pink bill, and streaking across the center of the breast.

white rump

short tail with white corners

♂ BREEDING

white wing bars

white marks on head

large, pinkish bill

thick streaks on underparts

♀

brown patches on back

streaked underparts

♂ NONBREEDING

bold, white wing patches

black head and back

rose-red breast

white belly

♂ BREEDING

VOICE High, sharp, explosive *sink* or *eeuk*, like squeak of sneakers on floor tiles; airy *vree* in flight; liquid, flute-like warble, slow and relaxed.
NESTING Loose, open cup or platform, usually in deciduous saplings, mid to high level; 2–5 eggs; 1–2 broods; May–Jul.
FEEDING Arthropods, fruit, seeds, and buds.
HABITAT Deciduous and mixed woods, parks, and orchards while breeding.
LENGTH 8in (20cm)
WINGSPAN 12½in (32cm)

Northern Cardinal

(S)

Cardinalis cardinalis

The male Northern Cardinal, or "redbird," is a familiar sight across southeastern Canada and the eastern US. Females are less showy, but have a prominent, reddish crest and red accents on its tan-colored outer tail and wing feathers. The male aggressively repels intruders and will occasionally attack his reflection in windows and various shiny surfaces. Northern Cardinals neither migrate nor molt to a duller plumage, making identification easy.

warm red overall ♂

reddish crest

buff-olive upperparts

red on outer tail feathers

dark patch not as extensive as male

grayish brown underparts

♀

darker bill

smaller, duller crest

brownish wings

◑

prominent crest

thick, oranged-red bill

bright red back and wings

black patch on face, extends onto throat

♂

long, red tail

brownish toes and legs

VOICE *Sharp, metallic tik, also bubbly chatters; loud, variable, sweet, slurred whistle,* tsee-ew-tsee-ew-whoit-whoit-whoit-whoit-whoit.
NESTING *Loose, flimsy cup of grass, bark, and leaves, in deciduous thicket; 2–4 eggs; 1–3 broods; Apr–Sep.*
FEEDING *Seeds and insects, such as beetles and caterpillars; also buds and fruit.*
HABITAT *Thickets of relatively moist habitats, such as deciduous woodland, scrub, desert washes, and backyards.*
LENGTH 8½in (22cm)
WINGSPAN 12in (30cm)

Indigo Bunting

S

Passerina cyanea

The brilliantly colored Indigo Bunting is actually a vibrant, almost cyan-blue, turning to indigo on the male's head before finally becoming a rich violet on the face. Indigo Buntings are specialists of disturbed habitats, originally depending on tree-falls within forests and the grassland-forest edge. Human activity, however, has radically increased suitable breeding habitats.

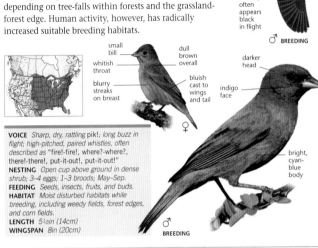

blue overall; often appears black in flight

♂ BREEDING

small bill

whitish throat

blurry streaks on breast

dull brown overall

bluish cast to wings and tail

♀

darker head

indigo face

bright, cyan-blue body

♂ BREEDING

VOICE *Sharp, dry, rattling pik!; long buzz in flight; high-pitched, paired whistles, often described as "fire!-fire!, where?-where?, there!-there!, put-it-out!, put-it-out!"*
NESTING *Open cup above ground in dense shrub; 3–4 eggs; 1–3 broods; May–Sep.*
FEEDING *Seeds, insects, fruits, and buds.*
HABITAT *Moist disturbed habitats while breeding, including weedy fields, forest edges, and corn fields.*
LENGTH *5½in (14cm)*
WINGSPAN *8in (20cm)*

Lazuli Bunting

S

Passerina amoena

Resembling a small bluebird, the Lazuli Bunting is closely related to the Indigo Bunting. Hybrids are locally common where their ranges overlap. Male hybrids are of two main types—the first resembles an Indigo Bunting with a white belly and wing bars, and the second resembles a dull Indigo Bunting with a brownish back. Females are harder to identify.

bold, white wing bars

♂ BREEDING

tawny breast

♀

blue tinge to wings and tail

unstreaked underparts

sky-blue head

mantle tinged with brown

dull orange breast

conspicuous white shoulder

white belly

sky-blue rump

♂

VOICE *Sharp, dry, rattling pik! similar to Indigo Bunting; higher, faster, thinner song with less repetition than Indigo Bunting.*
NESTING *Open cup above ground in dense shrub; 3–4 eggs; 1–2 broods; Apr–Aug.*
FEEDING *Seeds and fruits; insects while breeding.*
HABITAT *Open, disturbed habitats while breeding, especially alongside waterways and in thickets.*
LENGTH *5½in (14cm)*
WINGSPAN *8½in (22cm)*

Index

X

Acknowledgments

DORLING KINDERSLEY would like to thank the following contributors: David M. Bird, Nicholas L. Block, Peter Capainolo, Matthew Cormons, Malcolm Coulter, Joseph DiCostanzo, Shawneen Finnegan, Neil Fletcher, Ted Floyd, Jeff Groth, Paul Hess, Brian Hiller, Rob Hume, Thomas Brodie Johnson, Kevin T. Karlson, Stephen Kress, William Moskoff, Bill Pranty, Michael L. P. Retter, Noah Strycker, Paul Sweet, Rodger Titman, Elissa Wolfson, and Shramana Purkayastha for editorial assistance. **Map Editor** Paul Lehman.

The publisher would like to thank the following for their kind permission to reproduce their photographs:

Almost without exception, the birds featured in the profiles in this book were photographed in the wild.

(Key: a-above; b-below/bottom; c-center; f-far; l-left; r-right; t-top)

Alamy Images: Derrick Alderman 11cla; All Canada Photos 130br; Blickwinkel 11br; Rick & Nora Bowers 155t; Gay Bumgarner 11cr; Juniors Bildachiv 7ca.

Ardea: Peter Steyn 51bc.

Doug Backlund: 6-7ca, 72cra, 78cr, 218cb, 218crb.

Mike Danzenbaker: 51c, 52cra, 127cb, 130ca, 130cr, 131crb, 131br, 133cb, 152bc, 152clb, 154crb, 154br, 155clb, 155clb, 242crb, 259cra, 265ca.

DK Images: Chris Gomersall Photography 18c, 30c, 49cb, 49br, 64b, 71bl, 74cra, 74tc, 117crb, 118cb, 122crb, 128ca, 128cra, 128crb, 132cla, 135cra, 191cb, 214cb, 224ca, 224c; David Tipling Photo Library 31cr, 42br, 55cr, 108br, 114cr, 114cra, 117ca, 118ca, 120br, 135cr; Mark Hamblin 32cb, 35cra, 45cla, 68cla, 140cra, 149crb, 223tr; Mike Lane 18crb, 18br, 28crb, 35cl, 43br, 56cb, 65cl, 103cb, 107cr, 107br, 113cra, 118crb, 126crb, 129ca, 132cb, 191crb, 487bl, 488bl; Gordon Langsbury 92ca, 99ca; Tim Loseby 228cb; George McCarthy 21cl, 60bl, 63bl, 101bl, 123br, 126crb; Kim Taylor 182crb; Roger Tidman 26crb, 26br, 28cra, 37crb, 40c, 43cb, 45cr, 59ca, 65c, 101clb, 104br, 110cb, 111cra, 111cr, 125crb, 127ca, 129cr, 136ca, 223ca, 266cl; Steve Young 35br, 40cb, 43ca, 43cr, 52ca, 68cra,123cr, 123cr, 126br, 209cra, 266cb.

Dudley Edmondson:15ca, 17br, 23cl, 23crb, 55crb, 66cra, 72cb, 74cb, 75crb, 75cb, 79ca, 79cla, 88cb, 106br, 111bc, 112cla, 147crb, 188cb, 267ca, 267crb, 276cb.

Tom Ennis: 109cb.

Hanne & Jens Eriksen: 122cr, 127cr.

Neil Fletcher: 23tr, 25crb, 27crb, 28br, 214crb.

FLPA: Goetz Eichhorn/Foto Natura 46crb; S & D & K Maslowski 11cl; Geoff Moon; Winfried Wisniewski/Foto Natura 15cr.

Joe Fuhrman: 67c, 96crb.

Getty Images: Brad Sharp 11cra.

Bob Glover: 114cra, 120crb.

Melvin Grey: 58bc, 60tc, 60cra, 85cb, 91cla, 104ca, 103cra; Tom Grey: 71clb, 129crb, 136br, 190cb, 193cl, 258bc, 258crb, 261cla, 261c.

Martin Hale: 52clb.

Arto Juvonen: 20cb, 49cla, 49cra.

Kevin T. Karlson: 15clb, 39ca, 45cb, 77cb, 94cb, 95crb, 95cb, 109cb, 128cb, 247cb, 273ca.

Garth McElroy: 1c, 26tc, 38br, 39cr, 41cr, 46cl, 56cra, 55br, 60cb, 61c, 62cra, 63ca, 67cla, 67c, 83cb, 88cra, 91cr, 92crb, 93cr, 93crb, 94cra, 94crb, 98cra, 99cl, 101cra, 101tc, 102cra, 103cl, 106tc, 106crb, 115br, 116cr, 116cra, 119cra, 149cla, 149cb, 156ca, 167bc, 167crb, 171cl, 176cl, 179crb, 186cla, 187cb, 194ca, 194cla, 193cla, 195cla, 196cra, 197cra, 198cb, 200cra, 200br, 204c, 204t, 205br, 209bc, 210cb, 212cra, 215b, 217cr, 217cl, 219tc, 220cra, 221cra, 222ca, 225ca, 227c, 227cra, 230ca, 230cr, 233ca, 232cra, 232cb, 232crb, 233cb, 237crb, 239cra, 239crb, 239cb, 240cb, 241cl, 241cra, 245cla, 247crb, 246bc, 250cra, 253ca, 259ca, 262cb, 268crb, 269cla, 269br, 271ca, 272cb, 273cb, 275cb, 277cra, 276cra, 278cra, 279cra, 280ca, 280br, 283crb, 285crb.

Arthur Morris/Birds As Art: 122br.

Bob Moul: 56cl, 253cra.

Alan Murphy: 54cb, 134b, 153b, 159br, 160c, 163cb, 192cra, 195crb, 234cb, 238crb, 255b, 277cb, 281ca.

Tomi Muukonen: 80cla, 121br, 123crb.

naturepl.com: Barry Mansell 22c; Nigel Marven 130crb.

NHPA/Photoshot: Bill Coster 87bc; Dhritiman Mukherjee; Kevin Schafer 47cb. Wayne Nicholas: 86crb, 86bc.

Judd Patterson: 11, 17cra, 17tc, 77crb, 140cb, 256crb.

E. J. Peiker: 14cra, 24cra, 28cr, 29cb, 31cr, 31clb, 31bc, 33cra, 33cr, 33cb, 33br, 34ca, 34cr, 35cb, 42cr, 59cra, 71tc, 77cra, 79crb, 81t, 84crb, 89cb, 89tl, 90crb, 93ca, 95cr, 97cla, 100crb, 109cr, 111crb, 121crb, 124cla, 133crb, 136bc, 143cl, 156crb, 157cla, 161bc, 178bc, 183cla, 188cla, 189ca, 193c, 194bc, 199cla, 227clb, 231cr, 236bc, 243cr, 258cra, 260cb, 263cra, 279br.

Jari Peltomäki: 1c, 19ca, 20clb, 23cl, 44bc, 69bc, 80clb, 80crb, 84crb, 139b, 264crb.

Robert Royse: 14cb, 16cb, 19clb, 38cb, 57cra, 82clb, 90ca, 98tl, 98crb, 104cra, 106ca, 115crb, 122ca, 137bc, 138cra,176c, 207ca, 208cla, 215ca, 246ca, 251cb, 252cla, 265cra, 274cra, 274cb, 274crb, 277crb, 281cra.

Bill Schmoker: 24cb, 26cra, 27br, 46bc, 48crb, 50ca, 50cla, 50clb, 51bc, 59cb, 78clb, 79cb, 154cla, 154crb, 182ca.

Brian E. Small: 9fcra, 13cra, 13ca, 13cb, 13br, 14cr, 14br, 16bl, 18ca, 18cr, 25clca, 27ca, 27cr, 29cra, 29cl, 32br, 34cb, 34bl, 35br, 38cr, 41ca, 42ca, 45bc, 48cra, 66cl, 66cb, 70c, 70cla, 74clb, 75cra, 76cra, 82cla, 86cra, 90cb, 96cra, 96tc, 97tr, 98cb, 100cra, 99cb, 102br, 102cb, 104crb, 105ca, 105cra, 105br, 108ca, 112crb, 115cra, 116crb, 116br, 119bc, 119crb, 121ca, 124tc, 124crb, 135cra, 136cb, 138bc, 138clb, 140crb, 140crb, 141cra, 141cb, 143crb, 145c, 145cra, 144crb, 146cra, 147cra, 150bc, 152ca, 156br, 158cb, 160crb, 160cla, 162ca, 162crb, 163cra, 164cl, 164c, 165ca, 165cra, 166bc, 166crb, 168crb, 168cb, 169clb, 169cr, 171cra, 172ca, 172crb, 173cra, 173ca, 173crb, 174crb, 175cra, 175crb, 177clb, 178ca, 179br, 180cl, 180c, 181ca, 183cb, 184cra, 184cb, 185ca, 185crb,187cla, 186c, 187crb, 189cra, 190cra, 190ca, 197cb, 199crb, 201clb, 202ca, 202crb, 203ca, 203bc, 206cra, 207crb, 209crb, 210ca, 211cb, 211cla, 212crb, 213cra, 213tc, 216cra, 216cr..., 216cb, 218cra, 218tc, 219c..., 220br, 220cb, 221ca, 221cr..., 225cl, 225crb, 228ca, 227cr..., 229cb, 230bc, 231ca, 232c..., 235cra, 235bc, 235crb, 236c..., 237ca, 237cb, 238c, 240cra, ..., 241c, 247cra, 247ca, 242ca, ..., 242br, 243c, 244cra, 244ca, ..., 244bl, 245crb, 245cb, 246cra..., 248cb, 248ca, 249cb, 249br, ..., 251cra, 251crb, 252cra, 252c..., 253crb, 253cb, 254ca, 254br..., 256cra, 256ca, 257cl, 257cb, ..., 258ca, 258cb, 260ca, 260cla, ..., 262cra, 262crb, 263br, 264ca, ..., 265crb, 268c, 269cb, 270cra, ..., 270cb, 271cra, 271cb, 272cl, ..., 273cra, 273crb, 275cra, 275crl..., 279cra, 279cb, 280cra, 280cb, ..., 282cra, 282ca, 283cl, 283cra, 282cb, 282crb, 284cra, 284crb, 284cl, 285ca, 285cra, 285cb.

Bob Steele: 15ca, 16cra, 16cr, 20cla, 23cb, 24ca, 24br, 30ca, 41br, 48cla, 48clb, 54ca, 57cr, 57cb, 63cb, 66cb, 68clb, 68crb, 71cra, 75tc, 76ca, 77ca, 81b, 83cla, 84ca, 84cra, 85cla, 86ca, 88ca, 88bc, 89cl, 90cra, 91cb, 95ca, 97crb, 100tl, 100bl, 102ca, 105cb, 107ca, 108cla, 108cra, 108crb, 109ca, 110ca, 110cr, 113cra, 112c, 113crb, 113cb, 114crb, 114br, 117cl, 119ca, 120ca, 121cr, 124cb, 125cl, 127crb, 129cb, 133cra, 142c, 142cla, 144ca, 144cra, 146crb, 146cb, 147cla, 148c, 148cla, 151cla, 153ca, 156cra, 157c, 158ca, 162cra, 163cra, 163crb, 165crb, 165cb, 166ca, 166cra, 167ca,

167cra, 168ca, 168cra, 170b, 171crb, 174cra, 174ca, 177cb, 177crb, 179cra, 179cr, 181b, 182cra, 182cb, 185br, 189crb, 190crb, 191ca, 192b, 194crb, 196bc, 198ca, 198cla, 199cra, 199cb, 200cra, 201cr, 200crb, 202cr, 202br, 203cl, 203crb, 205cra, 206crb, 207cra, 207cb, 208c, 210br, 213crb, 214ca, 215ca, 216crb, 222cl, 222crb, 223cb, 223crb, 224crb, 226cb, 228cb, 229clb, 230crb, 231c, 231ca, 233br, 236crb, 237cra, 239c, 238clb, 243cla, 243clb, 248c, 252cb, 249tc, 254cra, 256crb, 256cb, 258br, 259crb, 259cb, 262ca, 263cr, 265cb, 267cra, 267cb, 268cl, 270ca, 271crb, 277ca, 278crb, 281crb.

Andy & Gill Swash: 61cla.

Glen Tepke: 131cra.

Markus Varesvuo: 4-5, 6-7bc, 12cb, 21crb, 32cra, 32cr, 37cra, 37cr, 37br, 41cb, 42crb, 46cra, 53bc, 55cra, 76cb, 76br, 80cra, 107crb, 126cr, 149cra, 229ca, 229cra, 264cb; Vireo: Robert L. Pitman 49bc, 131crb; Harold Stiver, 51fbl.

Peter S. Weber: 38cra, 61cr, 62c, 151cr, 162cb, 164cl.

Roger Wilmshurst: 3clb, 73c.

All other images © Dorling Kindersley
For further information see:
www.dkimages.com